Lecture Notes in Mathematics

Edited by A. Dold and B. Eckmann

1382

Hans Reiter

Metaplectic Groups and Segal Algebras

Springer-Verlag
Berlin Heidelberg New York London Paris Tokyo Hong Kong

Author

Hans Reiter
Institut für Mathematik der Universität
Strudlhofgasse 4, 1090 Vienna, Austria

Mathematics Subject Classification (1980): 43A70, 22D99, 11R56

ISBN 3-540-51417-1 Springer-Verlag Berlin Heidelberg New York
ISBN 0-387-51417-1 Springer-Verlag New York Berlin Heidelberg

Printing and binding: Druckhaus Beltz, Hemsbach/Bergstr.
2146/3140-543210 – Printed on acid-free paper

Herrn Professor André Weil

in Verehrung und Dankbarkeit

Preface

This work originated in 1976 in an attempt to understand the proof, and the nature, of the principal result, theorem 6, in A. Weil's memoir 'Sur certains groupes d'opérateurs unitaires' (Acta Math. 111, 143-211, 1964). But the scope soon widened when it became apparent that there is an intrinsic connexion between Weil's metaplectic group and a certain Segal algebra. Another task was to eliminate some difficulties of an analytical nature in Weil's work, hinted at by Weil himself in his Collected Papers (Springer, 1980), vol. III, p.445.

These notes may be considered as an addition to the author's book 'Classical Harmonic Analysis and Locally Compact Groups' (Oxford University Press, 1968). Weil's theory is taken up at the beginning and, as a rule, full proofs are provided, especially when the result has not been published before or differs, in some respect, from those in print. Ample references to the literature are given to facilitate the reader's task, especially with regard to comparison. Some new problems that arise are pointed out.

Some of the topics treated here were presented in seminars at the University of Vienna in 1978 and in 1982, but since then the whole matter has undergone profound changes.

A summary of the main results has appeared in a Note in the C.R. Acad. Sci. Paris, vol. 305 (I), 241-243 (1987), and I am grateful to Professor J. Dieudonné for presenting it to the Académie des Sciences. A short, non-technical survey of some underlying ideas is in the Proceedings of the 1987 Luxembourg Symposium on Harmonic Analysis (Springer LNM 1359, pp.238-241). I thank Professor J.-P. Pier cordially for the invitation to the symposium. I also wish to thank Dr. Marc Burger for permission to include here his strikingly simple unpublished proof of an important result due to J. Igusa.

Various difficulties have greatly delayed this work. It would not have been brought to completion without the friendly encouragement given at its inception by Professor C.L. Siegel; but now my grateful thanks come too late.

Of another - decisive - influence the work as a whole must speak.

February 1989 H. Reiter

Table of contents

Note. A reference such as § 2.6 means section 6 of § 2. Formulae are numbered anew in each section: § 2 (6-1) means formula 1 in § 2.6; but in § 2 itself it is simply referred to as (6-1). References are provided quite frequently and the reader may disregard them at will; but sometimes they may be helpful.

Introduction

The analytical aspects of Weil's memoir [25] form the main concern of these lectures, in particular his principal result (theorem 6) which is established here for a larger class of functions. This class, a so-called Segal algebra introduced by Feichtinger [7], seems to be of intrinsic interest in harmonic analysis, for a variety of reasons.

It is a traditional problem in harmonic analysis, once a result has been established for a certain class of functions, to find an 'appropriate' class for which the result holds. A classical example is the Riesz-Fischer theorem. In the present context we refer to Weil's own commentary on his work as to the choice of a class of functions for his results [27, vol. III, pp. 443-444 and 448].

A detailed summary of the contents follows.

§ 1 is devoted to some preliminary topics, and a rapid glance at sections 1.1-1.2 should suffice for a first reading. In particular, 1.2 contains a result from harmonic analysis, known for about 20 years [16, p. 121, formula (3)], the significance of which becomes clear in the present context. It will probably be found convenient to read sections 1.3-1.13 only when they are referred to later on. They contain some matters required for our purposes, but generally not readily available elsewhere.

Segal algebras are taken up in § 2. Their main properties are briefly described. These have already been discussed previously ([16, Ch. 6, § 2], [17]) and the relevant references are provided. Feichtinger [7] has introduced, even for general locally compact groups, a particular - and particularly important - Segal algebra which we treat here only for locally compact abelian groups G, and in a slightly different way suitable for applications to Weil's theory. We denote this Segal algebra by $\mathcal{S}^1(G)$. Weil himself in [25] considers the space $\mathcal{S}(G)$ of Schwartz-Bruhat functions. Poguntke [15] has shown that $\mathcal{S}(\mathbb{R}^n)$ belongs to $\mathcal{S}^1(\mathbb{R}^n)$, and Feichtinger [7] has extended this to $\mathcal{S}(G)$. Here we shall consider certain Banach spaces of functions on elementary groups $E \cong \mathbb{R}^n \times (\mathbb{R}/\mathbb{Z})^p \times \mathbb{Z}^q \times \Gamma_1$ (Γ_1 finite), with a finite number of derivatives, and shall show that they belong to $\mathcal{S}^1(E)$. The proof is a modification of Poguntke's method. This result has an immediate extension to general G. It follows that $\mathcal{S}(G)$ is a rather small part of

$\mathcal{S}^1(G)$ (but dense!). Following Losert [11], we then give a characterization of $\mathcal{S}^1(G)$ among Segal algebras by some functorial properties.

In § 3 we come to the beginnings of Weil's theory (of which no previous knowledge is assumed). Certain unitary operators in $L^2(G)$ introduced by Weil [25, p. 102] are discussed, in particular those associated to 'characters of the second degree'. It is remarkable that these Weil operators induce automorphisms of the Banach space $\mathcal{S}^1(G)$ [i.e. bounded, bijective linear operators of $\mathcal{S}^1(G)$], as Feichtinger [7] has shown. This is treated in detail, and products of Weil operators are investigated for later application to metaplectic groups. Let us mention that one important Weil operator can only be defined if G and its dual are isomorphic.

The basis has now been established for developing Weil's theory for functions in the Segal algebra $\mathcal{S}^1(G)$. Part of this task is taken up in § 4: various groups introduced by Weil are discussed, especially his group $\mathbb{B}(G)$ of unitary operators in $L^2(G)$ (which contains the Weil operators of § 3), and Weil's general symplectic group Sp(G). The first main result linking Weil's theory to $\mathcal{S}^1(G)$ is that <u>the operators of $\mathbb{B}(G)$ induce automorphisms of the Banach space</u> $\mathcal{S}^1(G)$. The proof is analogous to that of Weil for $\mathcal{S}(G)$; it uses the functorial properties of $\mathcal{S}^1(G)$ to the fullest extent. Then Weil's theta functions are discussed, and <u>the Poisson-Weil formula</u> (theorem 4 of [25]) <u>is proved for functions in</u> $\mathcal{S}^1(G)$, by a modification of Weil's method indicated in [20]. The case that $x \longmapsto 2x$ is an automorphism of G is given special consideration, as in Weil's memoir.

In § 5 vector spaces X of finite dimension over a local field $\mathbb{K}$ are discussed and a lemma on certain quadratic forms is stated which will be needed later. It can be proved very simply when $X = \mathbb{K}$; moreover, the cases char $\mathbb{K} = 2$ and char $\mathbb{K} \neq 2$ are quite different. A proof for the general case $X = \mathbb{K}^n$ is given in § 6, by induction; for char $\mathbb{K} = 2$ it is distinctly more difficult.

Weil's theory is then further developed in § 7 for the additive group of a vector space X as above. Here linear groups naturally arise as soon as the field $\mathbb{K}$ is taken into account (vector spaces over different local fields can have isomorphic additive groups). The fundamentals of harmonic analysis on X are briefly described, with references to the literature. The structure of Weil's 'characters of the second degree' on X is discussed: this is obvious when char $\mathbb{K} \neq 2$, but requires a more detailed discussion in the case char $\mathbb{K} = 2$. We apply Witt's theorem on the generators of the linear symplectic group to obtain, in a rather direct way, a result important in Weil's theory as developed here; this takes the place of Weil's theorem 1.

Weil's pseudosymplectic group Ps(X) and his metaplectic group Mp(X) can now be introduced in the local case (§ 8). The elements of Mp(X) define automorphisms of $\mathfrak{S}^1(X)$. The main result states that *the representation of* Mp(X) *in the Banach space* $\mathfrak{S}^1(X)$ *is continuous*. Here various results obtained in earlier sections find their application. In the case that char $\mathbb{K} \neq 2$, Mp(X) may be considered as a subgroup of $\mathrm{Aut}(L^2(X))$, and the result extends to a larger group.

Adele groups X_A and the main facts of harmonic analysis related to them are outlined in § 9, with references. The Segal algebra $\mathfrak{S}^1(X_A)$ is discussed. The metaplectic group in the adelic case is introduced in a way slightly different from, but equivalent to, Weil's method. It is shown that, *in the adelic case the metaplectic group has a representation in a certain subspace* $\mathfrak{S}_I(X_A)$ *of* $\mathfrak{S}^1(X_A)$ - with its own natural topology - *and this representation is continuous*. When the characteristic is not 2, there are some further results (as in the local case).

In § 10 the theorems in §§ 8, 9 are put together with a result on Segal algebras established in § 2 and yield at once, respectively, *Weil's theorem* 6 *for functions in* $\mathfrak{S}^1(X)$ *in the local case, and for functions in* $\mathfrak{S}_I(X_A)$ *in the adelic case*, thus for a rather larger class of functions than those of Schwartz-Bruhat.

Looking at the chain of proofs leading to the results in §§ 8-10, we may say, first, that they are free from considerations of the structure of locally compact abelian groups, and, secondly, that the case of characteristic 2 is rather difficult. Yet it is of interest to consider also this case, and to establish the results in the generality properly pertaining to them, in the spirit of Weil's memoir [25].

In the Appendix we prove Weil's theorem 1 which concerns general locally compact abelian groups (and is not needed for the proof of his theorem 6). Weil's method of proof in [25] - a modification of one first used by Segal [22] in a special case - is adapted to show some new aspects of certain Segal algebras of the type studied here.

This brings to completion a programme proposed in [20].

§1. *Preliminaries from harmonic analysis and group theory*

1.1 Here a brief survey of some concepts, notations and results used in the sequel is given; a detailed exposition may be found in the author's book [16]. Let G be a locally compact abelian (l.c.a.) group; we write the group law additively. Let H be a closed subgroup of G. We say that Haar measures dx, $d\xi$, $d\dot{x}$ on G, H, G/H, respectively, are canonically related or coherent if

$$d\xi\, d\dot{x} = dx, \text{ i.e. } \int_{G/H}\int_{H} f(x+\xi)\, d\xi\, d\dot{x} = \int_{G} f(x)\, dx, \quad f \in \mathcal{K}(G), \tag{1-1}$$

where $\mathcal{K}(G)$ is the space of complex-valued continuous functions on G with compact support; here

$$(T_H f)(\dot{x}) := \int_{H} f(x+\xi)\, d\xi, \quad \dot{x} = \pi(x) \quad [\pi\colon G \longrightarrow G/H]$$

is in $\mathcal{K}(G/H)$. Relation (1-1) extends, in appropriate form, to $f \in L^1(G)$, and we have

$$\|T_H f\|_1 \le \|f\|_1 \tag{1-2}$$

[16, Ch.3, §§ 4.1 - 4.7, pp.67 ff.]. We disregard, as usual, the distinction between L^p and $\mathcal{L}^p$, $p \ge 1$. We define $(L_y f)(x) := f(x-y)$.

A character of G is defined as a continuous function $\chi : G \longrightarrow T$ (multiplicative group of complex numbers of absolute value 1) such that $\chi(x+y) = \chi(x)\chi(y)$ for x, y in G. The dual group G* consists of all characters of G, with a familiar topology, and is again locally compact; writing $\chi(x) = \langle x, x^* \rangle$, we can also write the group law in G* additively: $\langle x, x_1^* + x_2^* \rangle := \langle x, x_1^* \rangle \langle x, x_2^* \rangle$. We often put $\chi_{a^*}(x) := \langle x, a^* \rangle$ for fixed $a^* \in G^*$; likewise we write $\chi_a(x^*) := \langle a, x^* \rangle$, which also serves to identify $(G^*)^*$ with G (duality theorem).

In practice, when G is given, G* will again be isomorphic to some 'concrete' group and may be identified with it; but the isomorphism is only determined up to an automorphism of this group. The concepts of isomorphism and automorphism will always be understood in the sense of topological groups, unless explicitly stated otherwise.

The Fourier transform (F.t.) of $\Phi \in L^1(G)$ is defined by

$$(\mathcal{F}_G\Phi)(x^*) \quad := \quad \hat{\Phi}(x^*) \quad := \quad \int_G \Phi(x)\overline{\langle x, x^*\rangle}\, dx$$

(note that this differs slightly from the definition used by Weil). It is also defined for $\Phi \in L^2(G)$, the two definitions being the same on $L^1 \cap L^2(G)$. We have $(L_y\Phi)^\wedge = \bar{\chi}_y \cdot \hat{\Phi}$ and $(\chi_{y*} \cdot \Phi)^\wedge = L_{y*}\hat{\Phi}$.

Let $\Lambda^1(G)$ be the subspace of all continuous functions Φ in $L^1(G)$ such that $\hat{\Phi} \in L^1(G^*)$. There is a dual Haar measure dx^* on G^* such that the following two theorems hold:

Inversion theorem. $\Phi(x) \;=\; \int_{G^*} \langle x, x^*\rangle \hat{\Phi}(x^*)\, dx^*, \quad x \in G$, for $\Phi \in \Lambda^1(G)$.

Plancherel's theorem. The mapping $\Phi \longmapsto \hat{\Phi}$ is a Hilbert space isomorphism of $L^2(G)$ onto $L^2(G^*)$, $\int_G |\Phi(x)|^2 dx \;=\; \int_{G^*} |\hat{\Phi}(x^*)|^2 dx^*$, $\Phi \in L^2(G)$.

Let $\Lambda^{\mathcal{K}}(G)$ consist of the continuous functions in $L^1(G)$ whose Fourier transforms have compact support; so $\Lambda^{\mathcal{K}}(G)$ is a subspace of $\Lambda^1(G)$. Let V^* be a compact neighbourhood (nd.) of 0 in G^*; we denote by $\Lambda^{a^*+V^*}$ the linear subspace of $\Lambda^{\mathcal{K}}(G)$ consisting of all $\Phi \in \Lambda^{\mathcal{K}}(G)$ with $\operatorname{Supp} \hat{\Phi} \subset a^*+V^*$ $(a^* \in G^*)$. Then we introduce

$$\mathfrak{S}^{V^*}(G) \quad := \quad \bigcup_{a^* \in G^*} \Lambda^{a^*+V^*}(G) \;=\; \left\{ \chi_{a^*} \cdot \Phi \;\middle|\; \Phi \in \Lambda^{V^*}(G),\; a^* \in G^* \right\}.$$

The set $\mathfrak{S}^{V^*}(G)$ is not a vector space (except in an obvious case), but admits multiplication by complex numbers and is invariant under translation, and under multiplication by a character. We may describe it as the set of all functions in $\Lambda^{\mathcal{K}}(G)$ whose Fourier transforms have support 'of width at most V^*'. It will play an important role in our work.

1.2 Given a closed subgroup H of G, one defines the orthogonal or associated subgroup H_* of G^* by

$$H_* \quad := \quad \left\{ \xi_* \;\middle|\; \xi_* \in G^*,\; \langle \xi, \xi_*\rangle = 1 \text{ for all } \xi \in H \right\}.$$

H is, by the identification of G^{**} with G, orthogonal to H_*. The group H_* is the dual of G/H. Fix dx on G; given $d\xi$ on H, determine $d\dot{x}$ on G/H by $dx = d\xi d\dot{x}$ and let $d\xi_*$ be the dual Haar measure on H_* : $d\xi_*$ is called the orthogonal measure to $d\xi$ (relatively to dx).

For any closed subgroup H (and coherent Haar measures) we have, if $\mathcal{F}_{G/H}$ denotes the Fourier transform on G/H:

$$\mathcal{F}_{G/H} T_H \Phi \quad = \quad R_{H_*} \hat{\Phi}\,, \qquad \Phi \in L^1(G),$$

R_{H_*} denoting the restriction to H_* (cf. [16, Ch. 4, § 4.3, p. 104]).

The following result will play a basic role later on:

Let H be a closed subgroup of G; fix Haar measures dx on G, $d\xi$ on H. Let V^* be a <u>compact</u> neighbourhood of 0 in G^*. There is a constant C such that

$$\int_H |\Phi(\xi)|\, d\xi \ \le\ C\cdot\int_G |\Phi(x)|\, dx \quad \text{for all } \Phi\in \mathfrak{S}^{V^*}(G). \tag{2-1}$$

This will be called the L^1-<u>restriction inequality for</u> $\mathfrak{S}^{V^*}(G)$. Here C may be chosen as follows. Let dx^* be the dual measure on G^*, $d\xi_*$ the orthogonal measure on H_*, and determine $d\dot{x}^*$ on G^*/H_* by coherence. Let g^*, h^* be functions in $\mathcal{K}(G^*)$ such that the convolution $g^* * h^*$ is 1 on V^*. Then we may take

$$C \ :=\ C_H(V^*) \ :=\ \|T_{H_*}(|g^*|)\|_2\cdot\|T_{H_*}(|h^*|)\|_2.$$

For the L^1-restriction inequality, we refer to [16, Ch.5, § 5.2, p.120 f.]; it is proved there (in dual form) for $\Lambda^{V^*}(G)$, but it is rather useful to state it for $\mathfrak{S}^{V^*}(G)$, an obvious extension. The inequality is used for the proof of <u>Poisson's formula</u> [16, Ch.5, § 5]:

$$\int_H \Phi(\xi)d\xi = \int_{H_*} \hat{\Phi}(\xi_*)d\xi_* \ \text{ for } \Phi\in\Lambda^{\mathcal{K}}(G) \quad (d\xi_* \text{ orthogonal to } d\xi). \tag{2-2}$$

By means of (2-2) we can show that <u>if the Haar measures on</u> G, H, G/H <u>are coherent, then so are the dual measures on</u> G^*, H_*, <u>and</u> $G^*/H_* = H^*$ [16, Ch.5, § 5.4, p.122].

The L^1-restriction inequality clearly implies: if $\Phi\in\Lambda^{\mathcal{K}}(G)$, then $T_H\Phi$ is continuous on G/H.

In this context, we also have: let Φ be in $\mathfrak{S}^{V^*}(G)$, $R_H\Phi$ its restriction to the closed subgroup H of G; then $R_H\Phi$ lies in $\mathfrak{S}^{V^{*\prime}}(H)$, where $V^{*\prime} := \pi_*(V^*)$ $[\pi_*: G^* \to G^*/H_*]$ [16, Ch. 5, § 5.5 (ii), p.122].

When H is open, the relation (2-1) is of course trivial, even for $\Phi\in L^1(G)$.

For an <u>open</u> subgroup $H\subset G$ - in which case H_* is compact - we note:

1. Let Φ be a continuous function in $L^1(G)$. If Φ vanishes outside H, then $\hat{\Phi}$ is H_*-periodic, and conversely (cf. [16, Ch.5, § 4.4 (ii), p.118] for a more general result).

2. For $\Phi\in L^1(G)$ and its restriction $R_H\Phi$ to H, we have $\mathcal{F}_H R_H\Phi = T_{H_*}\hat{\Phi}$ (this may be considered as a trivial case of Poisson's formula).

REMARK. Let Γ be a <u>discrete</u> subgroup of G, so that G^*/Γ_* is compact. Then there is a function $g\in\Lambda^{\mathcal{K}}(G)$ such that

$$\int_{\Gamma_*} \hat{g}(x^*+\gamma_*)\, d\gamma_* = 1 \quad \text{for all } x^* \in G^*.$$

For, take any $g_1^* \in \mathcal{K}(G^*)$ such that $T_{\Gamma_*} g_1^* = 1$ on G^*/Γ_*. Let $g_1 = \mathcal{F}_G^{-1} g_1^*$ and put $g = (g_1)^2$. Then $\hat{g} = g_1^* * g_1^*$, by Plancherel; moreover, $T_{\Gamma_*}\hat{g}$ is the convolution of the constant 1 on G^*/Γ_* with itself, thus 1 (we may suppose that G^*/Γ_* has Haar measure 1).

1.3 The following result will be used several times; thus we state it as a lemma, for reference.

LEMMA. Let G, G', H, be (separated) topological groups Let $f: G\times G' \longrightarrow H$, be continuous and such that, for all $x \in G$, $f(x,e') = e_H$, where e' is the neutral element of G', e_H that of H. Then, given a compact set K in G and a nd. V_H of e_H in H, there is a nd. V' of e' in G' such that $f(x,x') \in V_H$ for all $x \in K$ and all $x' \in V'$.

By the assumption on f, there is, for every $a \in K$, an open nd. $V(a)$ of a in G, and a nd. V'_a of e' in G' such that $f(x,x') \in V_H$ for $x \in V(a)$, $x' \in V'_a$. Since K is compact, there are finitely many points $a_n \in K$ such that the corresponding open nds. $V(a_n)$ cover K; we can put $V' = \bigcap_n V'_{a_n}$ to obtain the assertion.

1.4 Let G_1, G_2 be l.c.a. groups and F_0 a bicharacter on $G_1\times G_2$, i.e. a continuous function $G_1\times G_2 \longrightarrow T$ such that $x_1 \longmapsto F_0(x_1,x_2)$, $x_2 \longmapsto F_0(x_1,x_2)$ are characters of G_1, G_2, for fixed $x_2 \in G_2$, $x_1 \in G_1$, respectively. Then there is a morphism, (i.e., a continuous algebraic homomorphism) ρ_0 of G_2 into G_1^* such that

$$F_0(x_1,x_2) = \langle x_1, x_2\rho_0\rangle, \quad x_1 \in G_1,\ x_2 \in G_2 \tag{4-1}$$

(we write ρ_0 on the right, in analogy with row vectors and matrices). Conversely, if ρ_0 is given and F_0 is defined by (4.1), then F_0 is a bicharacter on $G_1\times G_2$.

Clearly ρ_0 is an algebraic homomorphism; thus we only have to show continuity at $x_2 = 0$. Since $F_0(x_1,0) = 1$ for all $x_1 \in G_1$, we can apply Lemma 1.3: given a compact set $K \in G_1$ and $\varepsilon > 0$, there is a nd. U of 0 in G_2 such that, if $x_2 \in U$, then $|F_0(x_1,x_2) - 1| < \varepsilon$ holds for all $x_1 \in K$ - i.e., $x_2\rho_0$ lies in a given nd. of 0 in G_1^*. The converse is obvious, since $(x_1,x_1^*) \longmapsto \langle x_1,x_1^*\rangle$ is a bicharacter of $G_1\times G_1^*$.

Let us change the notation slightly and write G, G' in place of G_1, G_2. If ρ is a morphism of G into G', then we consider the bicharacter $(x,x'^*) \longmapsto \langle x\rho, x'^*\rangle$ of $G\times G'^*$; by the above, there is a morphism ρ^* of G'^* into G^*, called the adjoint of ρ, such that

$$\langle x\rho, x'^* \rangle = \langle x, x'^*\rho^* \rangle, \quad x \in G, \quad x'^* \in G'^*. \tag{4-2}$$

If ρ *is an isomorphism* (as always, in the sense of *topological* groups) *of* G *onto* G', *then* ρ^* *is an isomorphism of* G'^* *onto* G^*. Indeed, we may then consider ρ^{-1} and $(\rho^{-1})^*$: we have $\langle x'\rho^{-1}, x^* \rangle = \langle x', x^*(\rho^{-1})^* \rangle$, $x' \in G'$, $x^* \in G^*$, and, putting $x' = x\rho$,

$$\langle x, x^* \rangle = \langle x\rho, x^*(\rho^{-1})^* \rangle = \langle x, x^*(\rho^{-1})^*\rho^* \rangle, \quad x \in G, \ x^* \in G^*,$$

thus $(\rho^*)^{-1}$ exists and

$$(\rho^*)^{-1} = (\rho^{-1})^* \tag{4-3}$$

is a morphism; so ρ^* is indeed an isomorphism. We simply write ρ^{*-1} for the terms in (4-3) ; ρ^{*-1} is an isomorphism of G^* onto G'^*.

If ρ_1 is a morphism of G into G', ρ_2 a morphism of G' onto G", then

$$(\rho_1\rho_2)^* = \rho_2^*\rho_1^*,$$

and, if ρ_1, ρ_2 are isomorphisms,

$$(\rho_1\rho_2)^{*-1} = \rho_1^{*-1}\rho_2^{*-1}.$$

In practice, the case $G' = G^*$ is of importance. We identify $(G^*)^*$ with G as in § 1.1; then ρ^* is also a morphism of G into G^* and, if we replace in (4-2) x by y and write x for x'^*, its defining property is

$$\langle x, y\rho \rangle = \langle y, x\rho^* \rangle, \quad x, \ y \text{ in } G.$$

If here $\rho^* = \rho$, then ρ is called *symmetric*.

Let Mor(G,G') be the additive group of all morphisms of G into G', Is(G,G') the set of all isomorphisms of G onto G' (which may be empty!), and Aut(G) the (multiplicative) group of all automorphisms of G (as a *topological* group). We note for later reference:

(i) If α = Mor(G,G), then $\alpha^* \in \mathrm{Mor}(G^*,G^*)$.

(ii) If $\beta \in \mathrm{Mor}(G,G^*)$, then $\beta^* \in \mathrm{Mor}(G,G^*)$; $\langle x, y\beta \rangle = \langle y, x\beta^* \rangle$ for x, y in G.

(iii) If $\gamma \in \mathrm{Mor}(G^*,G)$, then $\gamma^* \in \mathrm{Mor}(G^*,G)$; $\langle x^*\gamma, y^* \rangle = \langle y^*\gamma^*, x^* \rangle$ for x^*, y^* in G^*.

Further we note:

(iv) If $\alpha \in \mathrm{Aut}(G)$, then $\alpha^* \in \mathrm{Aut}(G^*)$, and $\alpha \longmapsto \alpha^{*-1}$ is an (algebraic) isomorphism of Aut(G) onto Aut(G^*).

(v) If $\beta \in \mathrm{Is}(G,G^*)$, then $\beta^* \in \mathrm{Is}(G,G^*)$, $\beta^{*-1} \in \mathrm{Is}(G^*,G)$, and

$$(\alpha\beta)^{*-1} = \alpha^{*-1}\beta^{*-1}, \quad \alpha \in \mathrm{Aut}(G).$$

(vi) If $\gamma \in \mathrm{Is}(G^*,G)$, then $\gamma^* \in \mathrm{Is}(G^*,G)$, $\gamma^{*-1} \in \mathrm{Is}(G,G^*)$, and

$$(\gamma\alpha)^{*-1} = \gamma^{*-1}\alpha^{*-1}, \quad \alpha \in \mathrm{Aut}(G).$$

(vii) If $x \longmapsto 2x$ is an automorphism of G, then $x^* \longmapsto 2x^*$ is the adjoint automorphism of G^*.

We shall later consider in particular the case of vector spaces [§ 7.1].

1.5 LEMMA. Suppose there exists an isomorphism ρ of G onto G' (so $\rho^* \in \mathrm{Is}(G'^*, G^*)$ [§ 1.4]). Let dx, dx' be Haar measures on G, G', and dx^*, dx'^* the dual Haar measures on G^*, G'^*, respectively. Then the relation $|\rho^*| = |\rho|$ holds.

Here the Haar moduli $|\rho|$, $|\rho^*|$ are defined by

$$\int_{G'} f(x')dx' = |\rho|\cdot\int_G f(x\rho)dx, \quad f \in L^1(G'),$$

$$\int_{G^*} g(x^*)dx^* = |\rho^*|\cdot\int_{G'^*} g(x'^*\rho^*)dx'^*, \quad g \in L^1(G^*).$$

REMARK. When G' = G, one takes dx' = dx; (so $|\rho|$ is independent of the choice of dx); then also $dx'^* = dx^*$. In the case $G' = G^*$, we take $dx' = dx^*$; then G'^* is identified with G, and $dx'^* = dx$.

The lemma can be proved by the inversion theorem. Let Φ be any function in $A^1(G)$; define Φ' on G' by

$$\Phi'(x') := \Phi(x'\rho^{-1}), \quad x' \in G'.$$

A simple calculation gives

$$(\Phi')^\wedge(x'^*) = |\rho|\cdot\hat{\Phi}(x'^*\rho^*). \tag{5-1}$$

Note that $\Phi' \in A^1(G')$. The inversion theorem for G says that

$$\Phi(x) = \int_{G^*} \langle x, x^*\rangle\hat{\Phi}(x^*)\, dx^*, \quad x \in G,$$

so, by the transformation $x^* = x'^*\rho^*$ and (5-1), (4-2) and the inversion theorem for G',

$$\Phi(x) = |\rho^*|\cdot|\rho|^{-1}\Phi'(x\rho) = |\rho^*|\cdot|\rho|^{-1}\cdot\Phi(x),$$

whence the result (Weil [25, § 2, Lemme 1, p.147]; for the case G' = G Braconnier [4, Ch.IV, Prop.8, p.83]).

1.6 In the group Aut(G) of (topological) automorphisms of a locally compact group G a natural topology can be defined so that Aut(G) becomes a topological group (Braconnier [4, Chap.IV, pp.56 ff.]). For the case at hand - G abelian, with additive group law - one defines the nds. $\mathcal{U}$ of the neutral element $e \in \mathrm{Aut}(G)$ by choosing any compact set $K \subset G$, any nd. V of 0 in G, and putting $\mathcal{U} = \mathcal{U}(K, V)$, where

$$\mathcal{U}(K, V) := \left\{\alpha \mid \alpha \in \mathrm{Aut}(G),\ x\alpha - x \in V,\ x\alpha^{-1} - x \in V \text{ for all } x \in K\right\}. \tag{6-1}$$

The nds $\mathcal{U}(K,V)$ form a basis at e. By definition, $\mathcal{U}$ is symmetric, $\mathcal{U}^{-1} = \mathcal{U}$; also

$$K\alpha \in K+V \text{ for } \alpha \in \mathcal{U}(K,V). \tag{6-2}$$

We recall that for $\alpha \in \mathrm{Aut}(G)$ the <u>Haar modulus</u> $|\alpha|$ is defined [§ 1.5, Remark] by

$$\int_G f(x)dx = |\alpha|\cdot\int_G f(x\alpha)dx, \quad f \in L^1(G). \tag{6-3}$$

REMARK 1. Sometimes it is useful to write $\mathrm{mod}_G(\alpha)$ in place of $|\alpha|$.

By (6.2) we have, if the Haar measure $m_G(K)$ is >0 and V is compact,

$$|\alpha| \leq m_G(K+V)/m_G(K), \quad \alpha \in \mathcal{U}(K,V).$$

It follows that the multiplicative function $\alpha \longmapsto |\alpha|$ on Aut(G) is continuous (Braconnier [4, Ch.IV, Prop.1, p.75]).

REMARK 2. If $\alpha \in \mathrm{Aut}(G)$ leaves the closed subgroup H of G invariant, then α induces an automorphism $\dot\alpha \in \mathrm{Aut}(G/H)$ and $|\alpha|_G = |\alpha|_H \cdot |\dot\alpha|_{G/H}$. This is readily proved by means of § 1.1 (cf. Braconnier [4, Ch.IV, Prop.6, p.83] for the general noncommutative case).

The nds. $\mathcal{U}$ may also be defined, equivalently, as follows. We put for any compact sets $K \subset G$, $K^* \subset G^*$, and any nd. U of 1 in T,

$$\mathcal{U} = \mathcal{U}(K,K^*,U) := \left\{\alpha \mid \alpha \in \mathrm{Aut}(G),\ \langle x\alpha - x, x^*\rangle \in U,\ \langle x\alpha^{-1} - x, x^*\rangle \in U \text{ for all } x \in K \text{ and all } x^* \in K^*\right\}.$$

If we introduce here

$$V = \left\{y \mid y \in G,\ \langle y, x^*\rangle \in U \text{ for all } x^* \in K^*\right\},$$

$$V^* = \left\{y^* \mid y^* \in G^*,\ \langle x, y^*\rangle \in U \text{ for all } x \in K\right\},$$

then V, V* are nds. of 0 in G, G*, respectively, and

$$\mathcal{U}(K,K^*,U) = \mathcal{U}(K,V).$$

Moreover, the relations

$$\langle x\alpha - x, x^*\rangle = \langle x, x^*\alpha^* - x^*\rangle, \quad \langle x\alpha^{-1} - x, x^*\rangle = \langle x, x^*\alpha^{*-1} - x^*\rangle$$

show that

$$\alpha \in \mathcal{U}(K,V) \iff \alpha^* \in \mathcal{U}'(K^*,V^*), \tag{6-4}$$

$\mathcal{U}'(K^*,V^*)$ being defined by

$$\mathcal{U}'(K^*,V^*) := \left\{\alpha^* \mid x^*\alpha^* - x^* \in V^*,\ x^*\alpha^{*-1} - x^* \in V^* \text{ for all } x^* \in K^*\right\}.$$

We now obtain:

PROPOSITION 1 (Braconnier [4, Ch.IV, Th.2, p.66]). <u>Let</u> G <u>be a locally compact abelian group</u>. <u>The mapping</u> $\alpha \longmapsto \alpha^{*-1}$ <u>is an</u>

<u>isomorphism of</u> Aut(G) <u>onto</u> Aut(G*) <u>in the sense of topological groups</u>.

We already know that this mapping is an isomorphism in the algebraic sense [§ 1.4 (iv)]. Now (6-4) shows that it is open; moreover (6-4) also shows that the nds. $\mathcal{U}(K,V)^{\alpha^{*-1}}$ form a nd. basis at e in Aut(G*), so the mapping is continuous.

We shall later consider the case G = X, a vector space of finite dimension over a local field (cf. § 5 and Braconnier [4, p.58]); the case of an 'elementary' abelian group is treated by Igusa [9, p.195].

For a general locally compact group G the topology of Aut(G) is defined by means of nds. $\mathcal{U}(K,V)$, where K is a compact set in G, V a nd. of the neutral element e in G:

$$\mathcal{U}(K,V) := \left\{ \alpha \mid \alpha \in \mathrm{Aut}(G),\ x^{\alpha}\cdot x^{-1} \in V,\ x^{\alpha^{-1}}\cdot x^{-1} \in V \text{ for all } x \in K \right\}.$$

We shall need the following result (cf. Braconnier [4], Ch.IV, Prop.1, p.59, and the remark at the end of footnote 4; cf. also the formulation given by Igusa [9, p.190] in the proof of his Prop.1):

PROPOSITION 2. <u>Let</u> $\mathbb{A}$ <u>be a locally compact group contained in a topological group</u> $\mathbb{B}$ <u>as a normal subgroup</u>. <u>For</u> $y \in \mathbb{B}$ <u>let</u> $\alpha(y) \in \mathrm{Aut}(\mathbb{B})$ <u>be defined by</u>

$$x \longmapsto x^{\alpha(y)} := y^{-1}xy,\ x \in \mathbb{A}.$$

<u>Then</u> α <u>is a continuous mapping of</u> $\mathbb{B}$ <u>into</u> Aut($\mathbb{A}$).

For the proof, let $f_1(x,y) := y^{-1}xyx^{-1}$, $f_2(x,y) := yxy^{-1}x^{-1}$, $x \in \mathbb{A}$, $y := \mathbb{B}$. The functions $f_j\colon \mathbb{A}\times\mathbb{B} \longrightarrow \mathbb{A}$ are continuous and $f_j(x,e) = e$ for all $x \in \mathbb{A}$; thus we can apply Lemma 1.3: given a compact set $K \subset \mathbb{A}$ and a nd. V of e in $\mathbb{A}$, there is a nd. V' of e in $\mathbb{B}$ such that $f_j(x,y) \in V$, j = 1, 2, for all $x \in K$ and all $y \in V'$. This means that

$$x^{\alpha(y)}\cdot x^{-1} \in V,\ x^{\alpha(y)^{-1}}\cdot x^{-1} \in V \text{ for all } x \in K, \text{ if } y \in V',$$

which gives the stated result.

<u>1.7</u> We introduce for $\alpha \in \mathrm{Aut}(G)$ the operator $M_1(\alpha)$ on $L^1(G)$ by

$$[M_1(\alpha)\Phi](x) := |\alpha|\cdot\Phi(x\alpha),\ x \in G,\ \Phi \in L^1(G), \tag{7-1}$$

where $|\alpha|$ is the <u>Haar modulus</u> of α (6-3); so, if dx is a (left) Haar measure on G,

$$\int_G [M_1(\alpha)\Phi](x)dx = \int_G \Phi(x)dx,\ \Phi \in L^1(G).$$

$M_1(\alpha)$ is an automorphism of the <u>algebra</u> $L^1(G)$, with convolution as multiplication; this fact is, implicitly, basic in the elementary exposition in [16, Ch. 1]. For G abelian, we note the formula

$$[M_1(\alpha)\Phi]^\wedge(x^*) = \hat{\Phi}(x^*\alpha^{*-1}), \quad \Phi \in L^1(G). \tag{7-2}$$

Hence, given any compact set $K^* \subset G^*$ and a compact nd. V^* of 0 in G^*, there is a nd. $\mathcal{U} = \mathcal{U}(K, V)$ of e in Aut(G) with the following property: if $\Phi \in L^1(G)$ and Supp $\hat{\Phi} \subset K^*$, then

$$\text{Supp } [M_1(\alpha)\Phi]^\wedge \subset K^*+V^* \text{ for all } \alpha \in \mathcal{U}.$$

Indeed, Supp $[M_1(\alpha)\Phi]^\wedge \subset K^*\alpha^*$ by (7-2); moreover, by § 1.6, Prop. 1, we can take $\mathcal{U}$ such that $\alpha^* \in \mathcal{U}'(K^*, V^*)$ for $\alpha \in \mathcal{U}$ (note that these nds. are symmetric). Then $x^*\alpha^*-x^* \in V^*$ for all $x^* \in K^*$ if $\alpha \in \mathcal{U}$, whence $K^*\alpha^* \subset K^*+V^*$ for $\alpha \in \mathcal{U}$.

We further note, for later reference,

$$\|M_1(\alpha)\Phi-\Phi\|_1 \longrightarrow 0, \quad \Phi \in L^1(G), \text{ when } \alpha \longrightarrow e \text{ in Aut}(G). \tag{7-3}$$

Clearly it is enough to show this for a function $u \in \mathcal{K}(G)$. We have

$$\|M_1(\alpha)u-u\|_1 \le \big||\alpha|-1\big| \cdot \int_G |u(x\alpha)|\,dx + \int_G |u(x\alpha)-u(x)|\,dx =$$
$$= \big|1-|\alpha|^{-1}\big| \cdot \|u\|_1 + \int_G |u(x\alpha)-u(x)|\,dx.$$

The function $\alpha \longrightarrow |\alpha|$ is continuous on Aut(G) [§ 1.6], and it is easy to see that the second term $\longrightarrow 0$ when $\alpha \longrightarrow e$ (cf. also Braconnier [4, Ch. IV, Prop. 2, p. 78]), whence (7-3).

<u>1.8</u> The Fourier transformation establishes an isomorphism of $L^2(G)$ onto $L^2(G^*)$; it also has an obvious close relation to the (unitary) operators of translation, and of multiplication by a character. In this context, we now discuss some groups introduced by Weil [25, n° 4]. We define for $u \in G$, $u^* \in G^*$,

$$\begin{aligned} [U(u,u^*)\Phi](x) &:= \Phi(x+u)\cdot\langle x, u^*\rangle, \quad \Phi \in L^2(G), \\ [U_*(u^*,u)\Phi'](x^*) &:= \Phi'(x^*+u^*)\cdot\langle u, x^*\rangle, \quad \Phi' \in L^2(G^*). \end{aligned} \tag{8-1}$$

The operators

$$t\cdot U(z), \quad z = (u, u^*) \in G\times G^*, \quad t \in T,$$

form a group $\mathbb{A}(G)$, <u>the 'Heisenberg group' attached to</u> G:

$$t_1\cdot U(z_1)\cdot t_2\cdot U(z_2) = t_1t_2F(z_1,z_2)\cdot U(z_1+z_2),$$

where

$$F(z_1,z_2) := \langle u_1, u_2^*\rangle \text{ for } z_j = \langle u_j, u_j^*\rangle, \; j = 1, 2. \tag{8-2}$$

Likewise the operators

$$t\cdot U_*(z_*), \quad z_* = (u^*, u) \in G^*\times G, \quad t \in T,$$

form a group $\mathbf{A}(G^*)$:

$$t_1\cdot U_*(z_{*1})\cdot t_2\cdot U_*(z_{*2}) = t_1t_2F_*(z_{*1},z_{*2})\cdot U_*(z_{*1}+z_{*2}),$$

where

$$F_*(z_{*1},z_{*2}) := \langle u_2,u_1^*\rangle \text{ for } z_{*j} = (u_j^*,u_j),\ j = 1,\ 2,$$

so that

$$F_*(z_{*1},z_{*2}) = F(z_2,z_1).$$

Thus $t\cdot U_*(z_*) \longmapsto t\cdot U(z)$ is an anti-isomorphism of $\mathbf{A}(G^*)$ onto $\mathbf{A}(G)$, while the relation

$$\mathcal{F}_G^{-1}t\cdot U_*(u^*,u)\mathcal{F}_G = t\cdot U(u,-u^*)$$

yields an isomorphism, as can be verified directly, of course.

With the strong topology for bounded operators in $L^2(G)$, $L^2(G^*)$, the groups $\mathbf{A}(G)$, $\mathbf{A}(G^*)$ are topological groups, and the isomorphism above is then in the sense of topological groups. For a Hilbert space $\mathcal{H}$, we define the <u>strong neighbourhoods</u> in $\mathcal{L}(\mathcal{H})$, the bounded linear operators on $\mathcal{H}$, as follows: we choose finitely many, say n, elements of $\mathcal{H}$, and $\varepsilon>0$, and put, if $\|\ \|$ denotes the norm in $\mathcal{H}$, for $A_o \in \mathcal{L}(\mathcal{H})$

$$\mathcal{U}(A_o) = \mathcal{U}_{\Phi_1,\dots,\Phi_n,\varepsilon}(A_o) := \left\{A \mid A\in \mathcal{L}(\mathcal{H}),\ \|A\Phi_j-A_o\Phi_j\| < \varepsilon,\ 1\le j\le n\right\}.$$

Let us now introduce the topological group A(G) of all pairs (z,t), $z\in G\times G^*$, $t\in T$, with the law of multiplication

$$(z_1,t_1)\cdot(z_2,t_2) = (z_1+z_2,F(z_1,z_2)\cdot t_1t_2)$$

and the product topology.

As Weil remarks [25, p. 149], $\mathbf{A}(G)$ <u>and</u> A(G) <u>are isomorphic as topological groups</u>. It will be worth while to give a proof here (cf. Igusa [9, p. 189, Lemma 1]).

The algebraic homomorphism $(u,u^*,t) \longmapsto t\cdot U(u,u^*)$ is an isomorphism, i.e. the kernel contains only the neutral element $(0,0,1)$: for, if $u \neq 0$, there is a $\Phi\in \mathcal{K}(G)$ such that $\Phi(0) = 1$, $\Phi(u) = 0$, so $[t\cdot U(u,u^*)\Phi](0) = 0$ and thus $\|t\cdot U(u,u^*)\Phi - \Phi\|_2 > 0$; likewise for u^*, by Plancherel, while the case of $t\cdot U(0,0)$ is trivial. Next the mapping is clearly continuous in the strong topology, since $t\cdot U(u,u^*) = t\cdot U(0,u^*)\cdot U(u,0)$. To show the continuity of the inverse, consider the neutral element $U(0,0)$ of $\mathbf{A}(G)$. Let nds. V, V^* of 0 in G, G^*, respectively, be given and let V_o be a nd. of 1 in T. Take $\Phi\in \mathcal{K}(G)$ such that $\|\Phi\|_2 = 1$. Let $\varepsilon>0$ be so small that $|t-1| < 2\varepsilon \Rightarrow t\in V_o$; there are nds. V_1, V_1^* of 0 in G, G* such that

$$\|U(u,u^*)\Phi - \Phi\|_2 < \varepsilon \quad \text{for } (u,u^*) \in V_1 \times V_1^*, \tag{8-3}$$

and we may assume $V_1 \subset V$, $V_1^* \subset V^*$. Next take $\Phi_1 \in \mathcal{K}(G)$ such that $\|\Phi_1\|_2 = 1$ and Supp $\Phi_1 \subset W$, a nd. of 0 in G for which $(-W)+W \subset V_1$. If the relation $\|t\cdot U(u,u^*)\Phi_1 - \Phi_1\|_2 < \sqrt{2}$ holds, then the functions $t\cdot U(u,u^*)\Phi_1$ and Φ_1 cannot have disjoint supports, so $u \in (-W)+W \subset V_1$. Likewise, take $\Phi_2 \in \Lambda^{\mathcal{K}}(G)$ such that $\|\Phi_2\|_2 = 1$ and Supp $\hat{\Phi}_2 \subset W^*$, a nd. of 0 in G^* satisfying $W^*+W^* \subset V_1^*$; by Plancherel and an application of the same argument to $\hat{\Phi}_2$, we obtain that $\|t\cdot U(u,u^*)\Phi_2 - \Phi_2\| < \sqrt{2}$ implies $u^* \subset V_1^*$. Thus, if $\|t\cdot U(u,u^*)\Phi_j - \Phi_j\|_2 < \varepsilon \le \sqrt{2}$, $j = 1, 2$, it follows that $u \in V_1 \subset V$, $u^* \in V_1^* \subset V^*$, and (8-3) holds. Moreover, if also, for the function Φ above, $\|t\cdot U(u,u^*)\Phi - \Phi\|_2 < \varepsilon$, then this and (8-3) yield $|t-1| = \|t\cdot\Phi - \Phi\|_2 < 2\varepsilon$, so $t \in V_0$. Thus the continuity of the inverse is established, which terminates the proof.

The centre of $\mathbf{A}(G)$ _consists of all operators_ $\Phi \longmapsto t\cdot\Phi$, $t \in T$, _on_ $L^2(G)$, as the law of multiplication shows; _analogously for_ $A(G)$.

As is well known, _the group_ $\mathbf{A}(G)$ _is irreducible_, i.e. there is no closed subspace of $L^2(G)$ invariant under $\mathbf{A}(G)$; for a short proof, cf. e.g. [18, § 4.4, p.353]; as is also well known, it follows that any bounded linear operator in $L^2(G)$ commuting with $\mathbf{A}(G)$ must be a constant multiple of the identity operator. In particular, _the centralizer of_ $\mathbf{A}(G)$ _in_ $\mathrm{Aut}(L^2(G))$, the unitary operators on $L^2(G)$, _coincides with the centre of_ $\mathbf{A}(G)$; a direct proof of this, due to Weil ([25], part of the proof of Th. 1) will be given in § 4.6.

_1.9_ Let us put $E_0 := \mathbb{R}^\nu \times (\mathbb{R}/\mathbb{Z})^p$. We call

$$\kappa := (\kappa_j)_{1\le j\le \nu+p}, \quad \lambda := (\lambda_k)_{1\le k\le \nu},$$

where κ_j, λ_k are integers ≥ 0, a _multi-index_; note that κ has 'length' $\nu+p$, while λ has length ν. We write $\kappa \le (m)$, where $m \ge 1$, to indicate that $\kappa_j \le m$ for $1 \le j \le \nu+p$; likewise, $\lambda \le (m)$ means: $\lambda_k \le m$ for $1 \le k \le \nu$.

An element $x \in E_0$ is of the form $x = (x', x'' \bmod \mathbb{Z}^p)$, with (x', x'') in $\mathbb{R}^\nu \times \mathbb{R}^p$; it is usually convenient, for shortness, to write simply $(x', x'') \in E_0$. A function on E_0 will always be represented by a function of $(x', x'') \in \mathbb{R}^\nu \times \mathbb{R}^p$, $\mathbb{Z}^p$-periodic in the variable x''.

For suitable functions Φ on E_0, we define the _differentiation operator_ D^κ, with κ as above, by

$$[D^\kappa\Phi](x) = [(\prod_{j=1}^{\nu+p} \{(2\pi i)^{-1}\partial/\partial x_j\}^{\kappa_j})\Phi](x), \quad x = (x_j)_{1\le j\le\nu+p} \in E_0.$$

We define the <u>multiplication operator</u> X^λ, with λ as above, by

$$[X^\lambda\Phi](x) = \prod_{k=1}^{\nu} x_k^{\lambda_k}\cdot\Phi(x), \quad x = (x_j)_{1\le j\le\nu+p}\in E_o.$$

For functions on the dual group $E_o^* := \mathbb{R}^\nu\times\mathbb{Z}^p$ we can define, correspondingly, the operators $X^{*\kappa}$ in the obvious way and, for suitable Φ^* on E_o^*, the operators $D^{*\lambda}$, with λ again as above, by

$$[D^{*\lambda}\Phi^*](x^*) = [(\prod_{k=1}^{\nu}\{(-2\pi i)^{-1}\partial/\partial x_k^*\}^{\lambda_k})\Phi^*](x^*), \quad x^* = (x_j^*)_{1\le j\le\nu+p}\in E_o^*.$$

We further introduce the <u>weight function</u>

$$w_o(x) := \prod_{j=1}^{\nu}(1+|x_j|), \quad x = (x_j)_{1\le j\le\nu+p} \in E_o, \tag{9-1}$$

which obviously satisfies

$$w_o(x+y) \le w_o(x)\cdot w_o(y), \quad x,\ y \text{ in } E_o. \tag{9-2}$$

<u>We define</u> $\mathcal{B}(E_o)$ as the space of all complex-valued functions Φ on E_o such that $D^\kappa\Phi$ exists and is continuous, and $w_o^3\cdot D^\kappa\Phi$ is bounded, for $\kappa \le (2)$; with the norm

$$\|\Phi\|_{E_o} := \max_{\kappa\le(2)} \|w_o^3\cdot D^\kappa\Phi\|_\infty, \quad \Phi\in\mathcal{B}(E_o),$$

this is a Banach space. It will have applications in § 2.

In this context, we shall use the following <u>Fourier relations for derivatives</u>: let Φ be a function on E_o, and $m \ge 1$.

(A) If $D^\kappa\Phi$ exists, is continuous, and $w_o^2\cdot D^\kappa\Phi$ is bounded, for $\kappa \le (m)$, then $\mathcal{F}(D^\kappa\Phi) = X^{*\kappa}\mathcal{F}\Phi$, and in particular $X^{*\kappa}\mathcal{F}\Phi$ is bounded, for $\kappa \le (m)$.

(A*) If Φ is continuous and $w_o^{m+2}\cdot\Phi$ is bounded, then for $\lambda \le (m)$ $D^{*\lambda}\mathcal{F}\Phi$ exists and $D^{*\lambda}\mathcal{F}\Phi = \mathcal{F}(X^\lambda\Phi)$, in particular $D^{*\lambda}\mathcal{F}\Phi$ is continuous.

Thus for functions in $\mathcal{B}(E_o)$ we have:

(B) If $\Phi\in\mathcal{B}(E_o)$, then $\mathcal{F}(D^\kappa X^\lambda\Phi) = X^{*\kappa}D^{*\lambda}\mathcal{F}\Phi$, $\kappa \le (2)$, $\lambda \le (1)$.

It will be noted that in the definition of $\mathcal{B}(E_o)$ the dimensions ν, p, do not matter (except, of course, for the group E_o itself).

In connexion with the space $\mathcal{B}(E_o)$, the following refinement of § 1.2, Remark, will be useful.

REMARK. Let $E_o^* = \mathbb{R}^\nu\times\mathbb{Z}^p$ be the dual of $E_o = \mathbb{R}^\nu\times(\mathbb{R}/\mathbb{Z})^p$, with the standard form of the duality relation. Let $\Gamma := \mathbb{Z}^\nu\times\{0\} \subset E_o$, so that $\Gamma_* = \mathbb{Z}^\nu\times\mathbb{Z}^p \subset E_o^*$, and let $V^* := [-1,1]^\nu\times\{0\}$, a compact nd. of 0 in E_o^*. <u>There is a function</u> g <u>on</u> E_o <u>with the following properties</u>:

(i) g is continuous and $w_o^3 \cdot g$ is bounded;
(ii) Supp $\hat{g}$ lies in V^*;
(iii) $\sum_{\gamma_* \in \Gamma_*} \hat{g}(x^* + \gamma_*) = 1$ for all $x^* \in E_o^*$.

Indeed, there is a function g_1^* on $\mathbb{R}$ such that $g_1^*(t) > 0$ for $|t| < 1$, $g_1^*(t) = 0$ for $|t| \geq 1$, g_1^* has a continuous third derivative on $\mathbb{R}$, and $\sum_{k \in \mathbb{Z}} g_1^*(t+k) = 1$ for all $t \in \mathbb{R}$; e.g. let $g_o^*(t) := (1-t^2)^4$ for $|t| \leq 1$, $g_o^*(t) := 0$ for $|t| \geq 1$, and put $g_1^*(t) := g_o^*(t) / \sum_{k \in \mathbb{Z}} g_o^*(t+k)$, $t \in \mathbb{R}$. Let $g_1 := \mathcal{F}_{\mathbb{R}}^{-1} g_1^*$. Then we can define g on E_o by

$$g := g_1^{\otimes \nu} \otimes 1^{\otimes p}, \text{ i.e. } g(x) := \prod_{j=1}^{\nu} g_1(x_j), \quad x = (x_j)_{1 \leq j \leq \nu+p} \in E_o.$$

1.10 We now extend the definition of $\mathcal{B}(E_o)$ given in § 1.9 to a group of the form

$$E := \mathbb{R}^\nu \times (\mathbb{R}/\mathbb{Z})^p \times \mathbb{Z}^q \times \Gamma_1 \quad (\nu, p, q \geq 0), \tag{10-1a}$$

where Γ_1 is a finite additive group; thus

$$E := E_o \times \Gamma, \text{ with } E_o := \mathbb{R}^\nu \times (\mathbb{R}/\mathbb{Z})^p, \; \Gamma := \mathbb{Z}^q \times \Gamma_1. \tag{10-1b}$$

A topological group isomorphic to E is called an elementary group. For a function Φ on E, we denote by Φ_γ, $\gamma \in \Gamma$, the function $x \longmapsto \Phi(x, \gamma)$, $x \in E_o$. We define the Banach space $\mathcal{B}(E)$ - which depends on the representation (10-1) - as the space of all functions Φ on E such that $\Phi_\gamma \in \mathcal{B}(E_o)$ for all $\gamma \in \Gamma$ and $\sum_{\gamma \in \Gamma} \|\Phi_\gamma\|_{E_o} < \infty$, with

$$\|\Phi\| := \sum_{\gamma \in \Gamma} \|\Phi_\gamma\|_{E_o}$$

as norm; thus $\mathcal{B}(E) = L^1(\Gamma, \mathcal{B}(E_o))$. We emphasize that $\mathcal{B}(E)$ has been defined by means of the representation (10-1) of E (which amounts to the choice of a co-ordinate system); we need not be concerned with obtaining a space invariant under all automorphisms of E - the motivation for the definition is rather to obtain a large space suitable for later applications [§ 2.15].

However, by a slight change in the definition, one can define a sequence of Banach spaces $\mathcal{B}_n(E)$, $n \geq 1$, of functions on E such that

(i) $\mathcal{B}_{n+1}(E) \subset \mathcal{B}_n(E) \subset \mathcal{B}(E)$, $n \geq 1$;
(ii) $\bigcap_{n \geq 1} \mathcal{B}_n(E) = \mathcal{S}(E)$, the space of Schwartz-Bruhat functions on E;
(iii) $\mathcal{B}_n(E)$ is invariant under automorphisms of E.

$\mathcal{B}_n(E)$ may be defined as follows: $\mathcal{B}_n(E)$ consists of all (complex-valued) functions Φ on E such that for all multi-indices $\kappa \leq (n+1)$

[§ 1.9] and the corresponding partial differentiation operators D^κ - defined for functions on E_0 in § 1.9 and extended to functions on E in the obvious way - we have: $D^\kappa\Phi$ exists, is continuous, and $w_E^{n+2}\cdot D^\kappa\Phi$ is bounded, where w_E is the weight function defined by

$$w_E(x) := 1 + \sum_{j=1}^{\nu} |x_{1j}| + \sum_{k=1}^{q} |x_{3k}|, \quad x = (x_1, x_2, x_3, x_4) \in E,$$

with E as in (10-1a). The norm in $\mathcal{B}_n(E)$ is defined by

$$\|\Phi\|_{(n)} := \max_{\kappa\le(n+1)} \sup_{x\in E} w_E(x)^{n+2}\cdot|D^\kappa\Phi(x)|.$$

Then $\mathcal{B}_n(E)$ is indeed a Banach space, and the above properties (i) and (ii) hold. To show (iii), we first note: the existence and continuity of $D^\kappa\Phi$ implies that of $D^\kappa(\Phi\circ\alpha)$ for $\alpha\in \mathrm{Aut}(E)$ [for, let E, E_0 and Γ be as in (10-1b); then every automorphism of E is of the form $(x', x'') \longmapsto (x'\alpha_{11}+x''\alpha_{21}, x''\alpha_{22})$, where $\alpha_{11} \in \mathrm{Aut}(E_0)$, $\alpha_{22} \in \mathrm{Aut}(\Gamma)$, $\alpha_{21} \in \mathrm{Mor}(\Gamma, E_0)$, and it can be shown that every automorphism of E_0 is induced, in the canonical way, by an automorphism of $\mathbb{R}^\nu\times\mathbb{R}^p$ leaving the subgroup $\{0\}\times\mathbb{Z}^p$ invariant]. Secondly we note that for $\alpha\in \mathrm{Aut}(E)$

$$\sup_{x\in E} w_E(x)^{n+2}\cdot|D^\kappa\Phi(x\alpha)| = \sup_{x\in E} w_E(x\alpha^{-1})^{n+2}\cdot|D^\kappa\Phi(x)|, \quad \kappa \le (n+1),$$

and that for some constant C_α

$$w_E(x\alpha^{-1}) \le C_\alpha\cdot w_E(x) \quad \text{for all } x\in E.$$

[We have $w_E(x) = \dot{w}(\dot{x})$, where $\dot{x}\in \mathbb{R}^\nu\times\mathbb{Z}^q$, the quotient of E by its maximal compact subgroup (which is invariant under α !); it can be shown that every $\dot{\alpha}\in \mathrm{Aut}(\mathbb{R}^\nu\times\mathbb{Z}^q)$ is the restriction of an automorphism of $\mathbb{R}^\nu\times\mathbb{R}^q$ leaving $\mathbb{R}^\nu\times\mathbb{Z}^q$ invariant, and the existence of C_α follows.]

<u>1.11</u> We shall need the following lemma in §§ 4.5, 4.6. For the terminology on integration theory, we refer to the short survey in [16, Ch.3, § 2, p.45 ff., and § 4.8, p.59].

LEMMA. <u>Let for</u> $j = 1,2$, μ_j <u>be a positive measure on the locally compact space</u> X_j <u>such that</u> $\mathrm{Supp}\,\mu_j = X_j$. <u>Let</u> $X := X_1\times X_2$ <u>and</u> $\mu := m_1\otimes\mu_2$, <u>so that</u> $\mathrm{Supp}\,\mu = X$. <u>Let</u> f_j <u>be a locally integrable function on</u> X_j, <u>vanishing outside a countable union of compact sets. Suppose there is a continuous function</u> Φ <u>on</u> X <u>such that</u> $\Phi(x_1,x_2) = f_1(x_1)\cdot f_2(x_2)$ μ-a.e. <u>on</u> X, <u>and neither</u> $f_1(x_1) = 0$ μ_1-a.e. <u>on</u> X_1 <u>nor</u> $f_2(x_2) = 0$ μ_2-a.e. <u>on</u> X_2. <u>Then there are CONTINUOUS functions</u> f'_j <u>on</u> X_j <u>such that</u> $f'_j(x_j) = f(x_j)$ μ_j-a.e. <u>on</u> X_j, $j = 1,2$, <u>and</u> $f'_1\otimes f'_2 = \Phi$.

For the proof, take first $u_2\in \mathcal{K}(X_2)$ such that $\int_{X_2} f_2u_2d\mu_2 \ne 0$. Then we have for any $u_1 \in \mathcal{K}(X_1)$, since $(f_1u_1)\otimes(f_2u_2)$ is in $L^1(X,\mu)$:

$$\int_{X_1} f_1u_1d\mu_1 \cdot \int_{X_2} f_2u_2d\mu_2 = \int_X (f_1u_1)\otimes(f_2u_2)d\mu = \int_X \Phi\cdot(u_1\otimes u_2)d\mu$$

or

$$C_2\cdot\int_{X_1} f_1u_1d\mu_1 = \int_{X_1} \Phi_1u_1d\mu_1,$$

where

$$C_2 := \int_{X_2} f_2u_2d\mu_2 \neq 0, \quad \Phi_1(x_1) := \int_{X_2} \Phi(x_1,x_2)u_2(x_2)d\mu_2, \quad x_1 \in X_1.$$

Here (as is readily seen) Φ_1 is continuous and vanishes outside a countable union of compact sets, since Φ has these properties. We put $f_1' = C_2^{-1}\cdot\Phi_1$; since $u_1 \in \mathcal{K}(X_1)$ was arbitrary, it follows that $f_1(x_1) = f_1'(x_1)$ μ_1-a.e. on X_1 (not just locally a.e.!). Likewise we obtain a continuous function f_2' on X_2, vanishing outside a countable union of compact sets, such that $f_2(x_2) = f_2'(x_2)$ μ_2-a.e. on X_2.

Clearly $f_1\otimes f_2$ and $f_1'\otimes f_2$ coincide μ-a.e. on X_1, and so do $f_1'\otimes f_2$ and $f_1'\otimes f_2'$; hence $f_1'\otimes f_2'$ coincides μ-a.e. on X with $f_1\otimes f_2$, hence with Φ, and thus (since Supp μ = X) $f_1'\otimes f_2' = \Phi$ which concludes the proof.

This lemma is implicit in Weil's memoir [25] on p.158, in the discussion of the mapping (e).

1.12 We formulate here an obvious remark which will be useful later. Let X, Y be topological spaces, f a continuous mapping of X into Y; let E_f be the graph of f in X×Y, i.e. $E_f := \{(x,f(x)) \mid x \in X\}$. Then the mapping $(x,f(x)) \longmapsto x$ is a homeomorphism of E_f (with the topology induced by the product topology) onto X.

1.13 Another remark used in the sequel is the following. Let H, H' be topological groups, π a morphism (:= continuous algebraic homomorphism) of H onto H'; let H_1 be the kernel of π. Suppose there is an open set A' in H' for which there is a continuous lifting ρ back to H, i.e. a continuous mapping ρ of A' into H such that $\pi(\rho(x')) = x'$ for all $x' \in A'$. Put $A = \pi^{-1}(A')$. Then A is homeomorphic to $H_1\times A'$ under the mapping $x \longmapsto (\xi,x')$, $x' := \pi(x)$, $\xi := x\cdot\rho(x')^{-1}$; also π is open, i.e. a strict morphism of H onto H'. Indeed, the mapping stated is continuous and surjective, with the continuous inverse $(\xi,x') \longmapsto \xi\cdot\rho(x')$. The openness of π then follows.

§2. *Segal algebras; the Segal algebra* $\mathfrak{S}^1(G)$

<u>2.1</u> We recall here the definition and the main properties of Segal algebras (cf. [16, Ch. 6, § 2, p. 126] and [17, § 4, p. 16]). We shall only consider Segal algebras for <u>abelian</u> locally compact groups (but the definition below also applies in the non-abelian case, with trifling changes). Fix a Haar measure on the l.c.a. group G. A linear subspace $S^1(G)$ of $L^1(G)$ is called a <u>Segal algebra</u> if it has the following properties:

1. $S^1(G)$ is dense in $L^1(G)$ and is invariant under translations.
2. $S^1(G)$ is a (complex) Banach space, say with norm $\| \ \|_S$, and for some constant C

$$\|\Phi\|_1 \le C\cdot\|\Phi\|_S \quad \text{for all } \Phi\in S^1(G),$$

i.e. the identity mapping of $S^1(G)$ into $L^1(G)$ is continuous. There is no loss of generality in assuming $C = 1$.

3. (i) The Segal norm is translation invariant:

$$\|L_y\Phi\|_S = \|\Phi\|_S \quad \text{for all } \Phi\in S^1(G),\ y\in G.$$

(ii) For each $\Phi\in S^1(G)$, the mapping $y \longmapsto L_y\Phi$ of G into $S^1(G)$ is continuous. It is enough to consider continuity at $0\in G$:

$$\|L_y\Phi - \Phi\|_S \longrightarrow 0 \text{ when } y \longrightarrow 0 \text{ in } G.$$

$S^1(G)$ has been defined as a <u>Banach space</u>, but it is, in fact, a <u>Banach algebra</u> under convolution, even an ideal in $L^1(G)$; for details, cf. the above references.

We list here the main properties of $S^1(G)$ which will be used later; for proofs, cf. again the references above.

(A) $S^1(G)$ <u>contains</u> $\Lambda^{\mathcal{K}}(G)$ (defined in § 1.1).

(B) $\Lambda^{\mathcal{K}}(G)$ <u>is dense in</u> $S^1(G)$. This follows from a stronger result proved in [16, Ch. 6, § 2.3, p. 128].

(C) For any <u>compact</u> set $K^*\subset G^*$, there is a constant $C_S(K^*)$ such that

$$\|\Phi\|_S \le C_S(K^*)\cdot\|\Phi\|_1 \quad \text{for all } \Phi\in S^1(G) \text{ for which Supp } \hat{\Phi}\subset K^*.$$

We may take $C_S(K) = \|\tau\|_S$, with $\tau\in\Lambda^{\mathcal{K}}(G)$ such that $\hat{\tau}$ is 1 on K^*.

REMARKS.

1. For <u>discrete</u> G, the only Segal algebra $S^1(G)$ is $L^1(G)$ itself. This follows at once from (A).

2. For <u>non-discrete</u> G, the intersection of all Segal algebras $S^1(G)$ is $\Lambda^{\mathcal{K}}(G)$ [17, § 5, (vii), p.26]. We note that, for non-discrete G, $\Lambda^{\mathcal{K}}(G)$ itself is not a Segal algebra, since then any Segal algebra $S^1(G)$ clearly contains functions Φ such that Supp $\hat{\Phi}$ is not compact.

3. The norm of a Segal algebra $S^1(G)$ is 'essentially' unique: if $\| \ \|$ is any other norm on $S^1(G)$ with the properties 2 and 3 above, then there is a constant C such that

$$C^{-1}\cdot\|\Phi\|_S \le \|\Phi\| \le C\cdot\|\Phi\|_S \quad \text{for all } \Phi\in S^1(G).$$

For, by Fourier transformation, we may consider $S^1(G)$ as a function algebra on the dual group G*; then we can apply a general result (cf., e.g., [16, Ch. 2, § 3.6, p.26]). But this is a pure existence result; to obtain such a constant C in a concrete case, use must be made of special information that may be available.

4. Let $S_1^1(G) \subset S_2^1(G)$, with Segal norms $\| \ \|_{(j)}$. Then for some constant C_1

$$\|\Phi\|_{(2)} \le C_1\cdot\|\Phi\|_{(1)} \quad \text{for all } \Phi\in S_1^1(G).$$

This is shown as in Remark 3.

<u>2.2</u> We say that $S^1(G)$ is an <u>invariant</u> <u>Segal</u> <u>algebra</u> if for every character χ of G we have for all $\Phi\in S^1(G)$: $\chi\cdot\Phi\in S^1(G)$ and $\|\chi\cdot\Phi\|_S = \|\Phi\|_S$, i.e. if $\Phi \longmapsto \chi\cdot\Phi$ is an isometry of $S^1(G)$ (in fact, a Banach algebra automorphism).

A Segal algebra may not admit multiplication by characters (this was shown by Cigler, cf. the reference in [17, § 3, (viii), p.26]), and even if it does, the Segal norm may not be invariant [16, Ch.6, § 2.8, p.128]. The concept of invariant Segal algebra was introduced by Feichtinger (with the terminology 'strongly character invariant' [7, p.270]); he showed the following property of an invariant Segal algebra [7, p.274]:

<u>Let</u> $S^1(G)$ <u>be</u> <u>an</u> <u>invariant</u> <u>Segal</u> <u>algebra</u>. <u>For</u> <u>every</u> $\Phi\in S^1(G)$ <u>the</u> <u>mapping</u> $a^* \longmapsto \chi_{a^*}\cdot\Phi$ <u>of</u> G* <u>into</u> $S^1(G)$ <u>is</u> <u>continuous</u>. It is enough to show that

$$\|\chi_{a^*}\cdot\Phi - \Phi\|_S \longrightarrow 0 \quad (a^* \to 0 \text{ in } G^*).$$

The proof is by 'reduction to $\Lambda^{\mathcal{K}}(G)$'. Consider first $f\in\Lambda^{\mathcal{K}}(G)$; let V* be a compact nd. of 0 in G* and let K* = Supp $\hat{f}$ + V*, a compact set. Since Supp $(\chi_{a^*}\cdot f)^\wedge \subset K^*$ for $a^* \in V^*$, we have [cf. § 2.1 (C)] for $C = C_S(K^*)$

$$\|\chi_{a^*}\cdot f - f\|_S \le C\cdot \|\chi_{a^*}\cdot f - f\|_1 \quad \text{for } a^* \in V^*,$$

hence $\|\chi_{a^*}\cdot f - f\|_S \to 0$ $(a^* \to 0$ in $G^*)$ for any $f \in \Lambda^{\mathcal{K}}(G)$. Now, given $\Phi \in S^1(G)$ and $\varepsilon > 0$, we can choose an $f \in \Lambda^{\mathcal{K}}(G)$ such that $\|\Phi - f\|_S < \varepsilon$ [§ 2.1 (B)]; then

$$\|\chi_{a^*}\cdot\Phi - \Phi\|_S \le \|\chi_{a^*}\cdot(\Phi - f)\|_S + \|\chi_{a^*}\cdot f - f\|_S + \|\Phi - f\|_S ,$$

which, by invariance, is

$$< 2\varepsilon + \|\chi_{a^*}\cdot f - f\|_S , \quad a^* \in G^*,$$

and hence

$$< 3\varepsilon \quad \text{for all } a^* \in U^*,$$

a suitable nd. of 0 in G^*. This proves the assertion.

An important <u>example of an invariant Segal algebra</u> is $\Lambda^1(G)$ - defined in § 1.1 - with the norm

$$\|f\|_{\Lambda^1} := \|f\|_1 + \|\hat{f}\|_1, \tag{2-1}$$

where we take as usual the dual Haar measure on G^*.

It follows at once that, <u>if</u> $S^1(G)$ <u>is an invariant Segal algebra, then the Heisenberg group</u> $\mathbb{A}(G)$ [§ 1.8] <u>acts on</u> $S^1(G)$ <u>as a group of Banach space automorphisms</u>, i.e., for $\Phi \in S^1(G)$ we have $t\cdot U(z)\Phi \in S^1(G)$, $\|t\cdot U(z)\Phi\|_S = \|\Phi\|_S$, and the mapping $t\cdot U(z) \longmapsto t\cdot U(z)\Phi$ of $\mathbb{A}(G)$ into $S^1(G)$ is continuous. Thus a connexion with Weil's theory appears here which will emerge more clearly later (§ 3).

<u>2.3 There exists a smallest invariant Segal algebra</u> $\mathfrak{S}^1(G)$, i.e. one contained in every invariant Segal algebra $S^1(G)$. This fundamental fact stands in marked contrast to the situation prevailing for general Segal algebras (cf. § 2.1, Remark 2); it was discovered by Feichtinger [7, Th.1, p.273]. His method and result even apply to general locally compact groups; but here only Segal algebras for abelian groups will be used, and for this case another approach, outlined in [20], will be given which has some technical advantages.

We have defined, for a given compact nd. V^* of 0 in G^*, a subset $\mathfrak{S}^{V^*}(G)$ of $L^1(G)$ [§ 1.1]. Now <u>let</u> $\mathfrak{S}^1(G)$ <u>consist of all</u> Φ <u>of the form</u>

$$\Phi = \sum_{n\ge1} \Phi_n , \quad \Phi_n \in \mathfrak{S}^{V^*}(G), \ n \ge 1, \quad \sum_{n\ge1} \|\Phi_n\|_1 < \infty . \tag{3-1}$$

Here the series $\sum_{n\ge1} \Phi_n$ converges also uniformly on G: for, by the inversion theorem, if $m_{G*}(V^*)$ denotes the dual Haar measure of V^*,

$$|f(x)| \le m_{G*}(V^*)\cdot\|\hat{f}\|_\infty \le m_{G*}(V^*)\cdot\|f\|_1 \quad \text{for } f\in \mathfrak{S}^{V^*}(G), \tag{3-2}$$

so that $\sum_{n\ge1}\Phi_n$ converges 'absolutely' in $\mathscr{C}^b(G)$, the space of bounded continuous functions on G, with the 'sup'-norm $\|\ \|_\infty$, or even in the subspace $\mathscr{C}^o(G)$ of continuous functions 'vanishing at infinity'. Thus Φ is in $L^1\cap\mathscr{C}^o(G)$.

Any series representation of Φ of the type (3-1) will be called a V^*-<u>representation of</u> Φ; we may call

$$\Phi = \sum_{n\ge1} f_n\cdot\chi_{a_n^*},\quad f_n\in\Lambda^{V^*}(G),\ a_n^*\in G^*,\ n\ge1,\quad \sum_{n\ge1}\|f_n\|_1<\infty,$$

a special V^*-representation.

Let us show that $\mathfrak{S}^1(G)$ <u>is a Segal algebra</u>. It is obviously a linear subspace of $L^1(G)$ invariant under translations, and contains for every $a^*\in G^*$ a function Φ_{a^*} such that $[\Phi_{a^*}]\hat{}(a^*)\ne0$; hence it is dense in $L^1(G)$, by a classical theorem of Wiener's (which will often be applied implicitly in the sequel!). We define a norm in $\mathfrak{S}^1(G)$ by

$$\|\Phi\|_{\mathfrak{S}} := \|\Phi\|_{V^*} := \inf\sum_{n\ge1}\|\Phi_n\|_1,\quad \Phi\in\mathfrak{S}^1(G), \tag{3-3}$$

where the infimum is taken over all V^*-representations (3-1) of Φ; clearly

$$\|\Phi\|_1 \le \|\Phi\|_{V^*},\quad \Phi\in\mathfrak{S}^1(G). \tag{3-4}$$

Thus $\mathfrak{S}^1(G)$ <u>is continuously embedded in</u> $L^1(G)$.

REMARK 1. For $\Phi\in\mathfrak{S}^{V^*}(G)$ equality holds in (3-4).

Further, by (3-2),

$$\|\Phi\|_\infty \le m_{G*}(V^*)\cdot\|\Phi\|_{V^*},\quad \Phi\in\mathfrak{S}^1(G), \tag{3-5}$$

thus $\mathfrak{S}^1(G)$ is also continuously embedded in $\mathscr{C}^o(G)$. With the norm (3-3) $\mathfrak{S}^1(G)$ is complete: it is enough to observe that, if $(\Phi^{(k)})_{k\ge1}$ is a sequence in $\mathfrak{S}^1(G)$ such that $\sum_{k\ge1}\|\Phi^{(k)}\|_{V^*}\le\infty$, then $\sum_{k\ge1}\Phi^{(k)}$ converges, by (3-5), in $\mathscr{C}^o(G)$, say to Φ, the function Φ is in $\mathfrak{S}^1(G)$, and the series converges to Φ in $\mathfrak{S}^1(G)$. The norm (3-3) is invariant under translation: for, every V^*-representation of Φ yields, by a translation L_y, a V^*-representation of $L_y\Phi$, which has the same 'sum' of L^1-norms, and conversely; thus $\|L_y\Phi\|_{V^*} = \|\Phi\|_{V^*}$, and it is similarly seen that $\|L_y\Phi-\Phi\|_{V^*}\to0$ $(y\to0)$. Likewise for $\chi_{a^*}\cdot\Phi$, $a^*\in G^*$; thus $\mathfrak{S}^1(G)$ <u>is invariant</u>.

REMARK 2. Let us note, for later application, that $\mathfrak{S}^1(G)$ <u>also lies</u>

in $L^2(G)$, since it consists of bounded functions in $L^1(G)$. Also

$$\|\Phi\|_2 \le m_{G*}(V*)^{1/2} \cdot \|\Phi\|_{V*}, \quad \Phi \in \mathfrak{S}^1(G), \tag{3-6}$$

i.e. $\mathfrak{S}^1(G)$ *is continuously embedded in* $L^2(G)$: indeed, for a representation (3-1) we have

$$\|\Phi_n\|_2 = \|\hat{\Phi}_n\|_2 \le m_{G*}(V*)^{1/2} \cdot \|\hat{\Phi}_n\|_\infty \le m_{G*}(V*)^{1/2} \cdot \|\Phi_n\|_1, \ n \ge 1,$$

whence (3-6). Moreover, $\mathfrak{S}^1(G)$ *is dense in* $L^2(G)$: this follows from § 2.1 (A).

In appearance, $\mathfrak{S}^1(G)$ depends on the nd. V* used in the definition (3-1), but in fact it is independent of the choice of V*; this will be shown below. If G is compact, we can take $V* = \{0\} \subset G*$; then $\mathfrak{S}^1(G)$ is the algebra, under convolution, of all 'absolutely convergent Fourier series' on G which, for $G := \mathbb{R}/2\pi\mathbb{Z}$, played such an important part in Wiener's classical work. For general l.c.a. G, $\mathfrak{S}^1(G)$ may be described as the algebra of all 'L^1-absolutely convergent series' of continuous functions in $L^1(G)$ that have Fourier transforms with support 'of width at most V* '. It turns out that $\mathfrak{S}^1(G)$ is a natural generalization of Wiener's classical algebra to general l.c.a. groups G - natural in the sense of functorial and other properties connected with Weil's work studied later. $\mathfrak{S}^1(G)$ may be called *the associated Segal algebra*: it is intrinsically related to G, as we shall see.

2.4 Let us now prove that $\mathfrak{S}^1(G)$ *is the smallest invariant Segal algebra*, i.e. contained in every other invariant Segal algebra $S^1(G)$ (Feichtinger [7, Th. 1, p. 273], with a slightly different notation). For, $S^1(G)$ certainly contains $\mathfrak{S}^{V*}(G)$ [§ 2.1(A)]; moreover, there is a constant $C_S(V*)$ such that [cf. § 2.1 (C)]

$$\|f\|_S \le C_S(V*) \cdot \|f\|_1 \quad \text{for all } f \in \Lambda^{V*}(G).$$

If $S^1(G)$ is invariant, then this holds for all $f \in \mathfrak{S}^{V*}(G)$. Now let Φ be in $\mathfrak{S}^1(G)$, $\Phi = \sum_{n\ge1} \Phi_n$ a V*-representation of Φ; then

$$\sum_{n\ge1} \|\Phi_n\|_S \le C_S(V*) \cdot \sum_{n\ge1} \|\Phi_n\|_1 ,$$

whence Φ is in $S^1(G)$, and even

$$\|\Phi\|_S \le C_S(V*) \cdot \|\Phi\|_{V*} . \tag{4-1}$$

Thus we have $\mathfrak{S}^1(G) \subset S^1(G)$, and (4-1) says that the embedding is continuous.

The minimal property above shows that $\mathfrak{S}^1(G)$ *is independent of the nd. V* used in its definition*: for, if W* is another compact nd. of 0 in G*, then $\mathfrak{S}^1_{V*}(G) \subset \mathfrak{S}^1_{W*}(G) \subset \mathfrak{S}^1_{V*}(G)$. We can say a little more about the norms: let $\tau \in A^{\mathcal{K}}(G)$ be such that $\hat{\tau}$ is 1 on $V^* \cup W^*$; then (4-1) yields

$$C^{-1} \cdot \|\Phi\|_{V^*} \le \|\Phi\|_{W^*} \le C \cdot \|\Phi\|_{V^*} , \quad \Phi \in \mathfrak{S}^1(G), \tag{4-2}$$

with $C = C_{V^*, W^*} := \max \left\{ \|\tau\|_{V^*}, \|\tau\|_{W^*} \right\}$.

The minimal property is, of course, of basic importance: in order to prove that an invariant Segal algebra coincides with $\mathfrak{S}^1(G)$, we only need to show that it is contained in $\mathfrak{S}^1(G)$!

2.5 (i) The L^1-restriction inequality for $\mathfrak{S}^{V^*}(G)$ in § 1 (2-1) has an application to $\mathfrak{S}^1(G)$, by the very definition of $\mathfrak{S}^1(G)$. Let H be a closed subgroup of G; then

$$\int_H |\Phi(\xi)| d\xi \le C_H(V^*) \cdot \|\Phi\|_{V^*} , \quad \Phi \in \mathfrak{S}^1(G), \tag{5-1}$$

the constant $C_H(V^*)$ being as in § 1.2 (for chosen Haar measures).

For, let $\Phi = \sum_{n \ge 1} \Phi_n$ be a V^*-representation (3-1); then, by the inequality mentioned,

$$\int_H |\Phi(\xi)| d\xi \le \sum_{n \ge 1} \int_H |\Phi_n(\xi)| d\xi \le C_H(V^*) \cdot \sum_{n \ge 1} \|\Phi_n\|_1,$$

and as this holds for all V^*-representations of Φ, (5-1) follows.

We call (5-1) the restriction inequality for $\mathfrak{S}^1(G)$; it will find application in § 2.8, and in § 10 in the context of metaplectic groups.

(ii) As another application of the L^1-restriction inequality § 1 (2-1) we prove here that Poisson's formula § 1 (2-2) holds for the functions in $\mathfrak{S}^1(G)$. Indeed, let Φ be in $\mathfrak{S}^1(G)$ and $\Phi = \sum_{n \ge 1} \Phi_n$ a V^*-representation of Φ. Then $\sum_{n \ge 1} \int_H |\Phi_n(\xi)| d\xi < \infty$ by § 1 (2-1) and thus

$$\int_H \Phi(\xi) d\xi = \sum_{n \ge 1} \int_H \Phi_n(\xi) d\xi ,$$

which by Poisson's formula for $\Phi_n \in \mathfrak{S}^{V^*}(G)$ [cf. § 1 (2-2]) is

$$= \sum_{n \ge 1} \int_{H_*} \hat{\Phi}_n(\xi_*) d\xi_* ,$$

and this in turn is

$$= \int_{H_*} \sum_{n \ge 1} \hat{\Phi}_n(\xi_*) d\xi_* = \int_{H_*} \hat{\Phi}(\xi_*) d\xi_* ,$$

since $\sum_{n \ge 1} \int_{H_*} |\hat{\Phi}_n(\xi_*)| d\xi_* < \infty$, as the following argument shows: each Φ_n is in $\mathfrak{S}^{V^*}(G)$, thus $\operatorname{Supp} \hat{\Phi}_n \subset a_n^* + V^*$ for some $a_n^* \in G^*$, hence

$$\int_{H_*} |\hat{\Phi}_n(\xi_*)|d\xi_* \le m_{H_*}((a_n^*+V^*)\cap H_*)\cdot \|\hat{\Phi}_n\|_\infty \le C\cdot\|\Phi_n\|_1, \text{ for } n \ge 1,$$

where

$$C := \sup_{x^*\in G^*} m_{H_*}((x^*+V^*)\cap H_*) \le \max\{[T_{H_*}f^*](\dot{x}^*) \mid \dot{x}^*\in G^*/H_*\} < \infty,$$

f^* being any function in $\mathcal{K}_+(G^*)$ which is 1 on V^*.

<u>2.6</u> We discuss here those functorial properties of $\mathcal{G}^1(G)$ which have applications to Weil's theory treated in these lectures.

(i) <u>Let</u> ρ <u>be an isomorphism of</u> G <u>onto</u> G'. <u>Then</u>

$$\Phi \longmapsto \Phi', \quad \Phi\in \mathcal{G}^1(G), \quad \Phi' := |\rho|^{-1}\cdot(\Phi\circ\rho^{-1}),$$

<u>is an isomorphism of</u> $\mathcal{G}^1(G)$ <u>onto</u> $\mathcal{G}^1(G')$ [for $|\rho|$ cf. § 1.5]; <u>if</u> V^* <u>is a compact neighbourhood of</u> 0 <u>in</u> G^*, <u>then</u>

$$\|\Phi'\|_{V'^*} = \|\Phi\|_{V^*}, \quad \text{where } V'^* := V^*\rho^{*-1} \in G'^*.$$

In particular, <u>an automorphism of</u> G <u>induces an automorphism of</u> $\mathcal{G}^1(G)$ (cf. Feichtinger [7, Th. 1, p.273]).

If Φ has the V^*-representation $\Phi = \sum_{n\ge1}\Phi_n$, then $\Phi' = \sum_{n\ge1}\Phi'_n$ is a V'^*-representation of Φ', since $\mathcal{F}_{G'}\Phi'_n = \hat{\Phi}_n\circ\rho^*$ and $\sum_{n\ge1}\|\Phi'_n\|_1 = \sum_{n\ge1}\|\Phi_n\|_1$; conversely, every V'^*-representation of Φ' gives a V^*-representation of Φ, with the same 'sum' of L^1-norms. Thus $\Phi \longmapsto \Phi'$ is a bijection of $\mathcal{G}^1(G)$ onto $\mathcal{G}^1(G')$ which preserves the norm, and also an algebraic isomorphism.

(ii) Let $G = G_1\times G_2$; then $\mathcal{G}^1(G_1\times G_2)$ has the following <u>Segal tensor representation</u> in terms of $\mathcal{G}^1(G_1)$ and $\mathcal{G}^1(G_2)$: $\mathcal{G}^1(G_1\times G_2)$ <u>consists of all functions</u> Φ <u>on</u> $G_1\times G_2$ <u>of the form</u>

$$\Phi = \sum_{n\ge1}\Phi_{1n}\otimes\Phi_{2n}, \quad \Phi_{jn}\in\mathcal{G}^1(G_j),\ j = 1, 2, \quad \sum_{n\ge1}\|\Phi_{1n}\|_{(1)}\cdot\|\Phi_{2n}\|_{(2)} < \infty. \tag{6-1}$$

Here the tensor product of functions Φ_j on G_j is defined by

$$[\Phi_1\otimes\Phi_2](x_1,x_2) := \Phi_1(x_1)\cdot\Phi_2(x_2), \quad (x_1,x_2)\in G_1\times G_2,$$

and $\|\ \|_{(j)}$ denotes a norm in $\mathcal{G}^1(G_j)$. The series for Φ in (6-1) converges 'absolutely' in $\mathcal{C}^0(G_1\times G_2)$, since $\|\Phi_{jn}\|_\infty \le C_j\cdot\|\Phi_{jn}\|_{(j)}$ [cf. (3-5)]. <u>The norm in</u> $\mathcal{G}^1(G_1\times G_2)$ <u>can be defined by the</u> '<u>tensor norm</u>'

$$\|\Phi\|_{\mathcal{G}} := \inf\sum_{n\ge1}\|\Phi_{1n}\|_{(1)}\cdot\|\Phi_{2n}\|_{(2)}, \tag{6-2}$$

the infimum being taken for all Segal tensor representations of Φ (Feichtinger [7, Th.7, D, p.281]. <u>When</u> G <u>is the product of a finite</u>

<u>number of groups, the analogues of</u> (6-1) <u>and</u> (6-2) <u>likewise hold for</u> $\mathcal{S}^1(G)$.

It is readily seen that the space of functions (6-1), with (6-2) as a norm, is a Segal algebra which is invariant. To show that this is contained in $\mathcal{S}^1(G_1 \times G_2)$ - hence equal to it -, choose compact nds. V_j^* of 0 in G_j^* to fix the norms in $\mathcal{S}^1(G_j)$. Let Φ be a function on $G_1 \times G_2$ of type (6-1) above; each Φ_{jn} in (6-1) has a V_j^*-representation in $\mathcal{S}^1(G_j)$:

$$\Phi_{jn} = \sum_{k\geq 1} \Phi_{nk}^{(j)}, \quad \Phi_{nk}^{(j)} \in \mathcal{S}^{V_j^*}(G_j) \text{ , with } \sum_{k\geq 1} \|\Phi_{nk}^{(j)}\|_1 \leq 2\cdot\|\Phi_{jn}\|_{(j)},$$

say. From this we obtain, by rearranging a (pointwise absolutely convergent) double series of functions on $G_1 \times G_2$ into a single series,

$$\Phi = \sum_{m\geq 1} \Phi_m, \quad \sum_{m\geq 1} \|\Phi_m\|_1 < \infty ,$$

and this is clearly a V^*-representation with $V^* := V_1^* \times V_2^*$; thus Φ is in $\mathcal{S}^1(G_1 \times G_2)$.- For a product of <u>finitely</u> many groups G_j the proof is the same (<u>not</u> by induction!).

When G is a product of finitely many groups, we thus have two 'natural' norms in $\mathcal{S}^1(G)$. The norm in a Segal algebra is 'essentially' unique, in the sense of § 2.1, Remark 3; but we have no method for obtaining a constant relating the two norms, as mentioned there.

(iii) <u>The Fourier transformation</u> $\Phi \longmapsto \hat{\Phi}$, $\Phi \in \mathcal{S}^1(G)$, <u>yields a Banach space isomorphism of</u> $\mathcal{S}^1(G)$ <u>onto</u> $\mathcal{S}^1(G^*)$ (Feichtinger [7, Th.7, A, p.281]).

First we note that, when $\Phi \in \mathcal{S}^1(G)$, then $\hat{\Phi} \in L^1(G^*)$, i.e. $\mathcal{S}^1(G)$ lies in $A^1(G)$ [cf. (2-1) and use the final part of the argument in § 2.5 (ii) with $H_* := G^*$]. It is now easy to verify that $\mathcal{F}_G \mathcal{S}^1(G)$, with the norm $\|\hat{\Phi}\| := \|\Phi\|_{\mathcal{S}}$, is a Segal algebra (for the density property, Wiener's theorem can be used); it is, of course, invariant, thus $\mathcal{S}^1(G^*) \subset \mathcal{F}_G \mathcal{S}^1(G)$. The same argument, applied to $\mathcal{S}^1(G^*)$ and $\mathcal{F}_G^{-1}$, yields the reverse inclusion, hence equality. The 'transported' norm for $\mathcal{S}^1(G^*)$ is equivalent to the standard norm in $\mathcal{S}^1(G^*)$, by a general result [§ 2.1, Remark 3]; but we have no means for finding a constant relating the norms, as stated there. This is analogous to the situation in (ii).

(iv) <u>Let</u> $G = G_1 \times G_2$. <u>The partial Fourier transformation</u> $\Phi \longmapsto \mathcal{F}_2\Phi$,

$$[\mathcal{F}_2\Phi](x_1, x_2^*) := \int_{G_2} \Phi(x_1, x_2)\overline{\langle x_2, x_2^* \rangle} dx_2, \quad \Phi \in \mathcal{S}^1(G_1 \times G_2),$$

<u>yields a Banach space isomorphism of</u> $\mathcal{S}^1(G_1 \times G_2)$ <u>onto</u> $\mathcal{S}^1(G_1 \times G_2^*)$

(Feichtinger [7, Remark 17, p.283]).

This is a combination of the tensor representation (ii) with (iii) and the analogous tensor representation of $\mathfrak{S}^1(G_1 \times G_2^*)$.

2.7 We prove here, for later application, a property naturally connected with the tensor representation § 2.6 (ii). Let $G = \prod_{j=1}^{n} G_j$. The mapping

$$(\Phi_j)_{1\le j\le n} \longmapsto \Phi := \bigotimes_{j=1}^{n} \Phi_j \, , \quad \Phi \in \mathfrak{S}^1(G_j), \quad 1 \le j \le n,$$

of $\prod_{j=1}^{n} \mathfrak{S}^1(G_j)$ into $\mathfrak{S}^1(G)$ is continuous.

We choose the norm $\| \;\|_{(j)}$ in each $\mathfrak{S}^1(G_j)$ and let $\| \;\|_{\mathfrak{S}}$ be the resulting tensor norm in $\mathfrak{S}^1(G)$. Let $\Phi'_j \in \mathfrak{S}^1(G_j)$, $1 \le j \le n$, and put $\Phi' := \bigotimes_{j=1}^{n} \Phi'_j$. Then we can write $\Phi - \Phi'$ in the form

$$\Phi - \Phi' = \sum_{i=1}^{n} (\Psi_i - \Psi_{i-1}) = \sum_{i=1}^{n} \Psi'_i \, , \tag{7-1}$$

where $\Psi_i \in \mathfrak{S}^1(G)$ is defined by

$$\Psi_i := \bigotimes_{j=1}^{n} \Psi_{ij} \, , \quad \Psi_{ij} := \begin{cases} \Phi_j & \text{if } j \le i \\ \Phi'_j & \text{if } j > i \end{cases} , \quad 0 \le i \le n,$$

so that $\Psi_0 = \Phi'$, $\Psi_n = \Phi$, and $\Psi'_i \in \mathfrak{S}^1(G)$ is defined by

$$\Psi'_i := \Psi_i - \Psi_{i-1} = \bigotimes_{j=1}^{n} \Psi'_{ij}, \quad 1 \le i \le n, \quad \Psi'_{ij} := \begin{cases} \Phi_j & \text{if } j < i \\ \Phi_i - \Phi'_i & \text{if } j = i, \\ \Phi'_j & \text{if } j > i \end{cases}$$

$$\text{where } 1 \le j \le n. \tag{7-2}$$

Thus (7-1) is a Segal tensor representation of $\Phi - \Phi' \in \mathfrak{S}^1(G)$. Now let, for simplicity,

$$\|\Phi_j - \Phi'_j\|_{(j)} \le 1, \; 1 \le j \le n, \; M := \max\left\{ 1 + \|\Phi'_j\|_{(j)} \;\middle|\; 1 \le j \le n\right\}.$$

Then we have for the tensor norm in $\mathfrak{S}^1(G)$, by (7-1) and (7-2), the (gross) estimate

$$\|\Phi - \Phi'\|_{\mathfrak{S}} \le M^{n-1} \cdot \sum_{j=1}^{n} \|\Phi_j - \Phi'_j\|_{(j)} \longrightarrow 0 \quad (\Phi_j \longrightarrow \Phi'_j, \; 1 \le j \le n),$$

as was to be shown.

2.8 We establish here two functorial properties of $\mathfrak{S}^1(G)$ which are of intrinsic interest and will also be used later (for another application to Weil's theory cf. the proof of the generalization of the

Weil-Cartier theorem [19], where $\mathcal{S}^1(G)$ may be used instead of $\mathcal{S}(G)$).

(i) Let H be any closed subgroup of G. The restriction mapping

$$R_H: \quad R_H\Phi := \Phi|H, \quad \Phi\in\mathcal{S}^1(G),$$

is a Banach space morphism of $\mathcal{S}^1(G)$ onto $\mathcal{S}^1(H)$; in particular, $R_H\Phi$ is in $\mathcal{S}^1(H)$ if Φ is in $\mathcal{S}^1(G)$. More explicitly, choose Haar measures on G and H and let $\|\ \|_{V*}$ be the norm in $\mathcal{S}^1(G)$ determined by a compact nd. V* of 0 in G*; then the norm in $\mathcal{S}^1(H)$ corresponding to π_*V^* [π_*: $G^* \to G^*/H_*$] satisfies

$$\|R_{H*}\Phi\|_{\pi_*V^*} \le C_H(V^*)\cdot\|\Phi\|_{V^*}, \quad \Phi\in\mathcal{S}^1(G), \tag{8-1}$$

with the constant $C_H(V^*)$ as in § 1.2 (cf. Feichtinger, [7, Th.7, C, p.281]).

By the restriction inequality for $\mathcal{S}^1(G)$ [(5-1)], $R_H\Phi$ is in $L^1(H)$ for all $\Phi\in\mathcal{S}^1(G)$. It is easy to see that $R_H\mathcal{S}^1(G)$, with the quotient norm, is an invariant Segal algebra; thus we need only show that it lies in $\mathcal{S}^1(H)$. Now for $f\in\mathcal{S}^{V^*}(G)$ we have $R_Hf\in\mathcal{S}^{V^{*\prime}}(H)$, where $V^{*\prime} := \pi_*V^*$ [§ 1.2]; it readily follows [cf. § 1, (2-1)]: a V*-representation of $\Phi\in\mathcal{S}^1(G)$ yields a V*'-representation of $R_H\Phi$, i.e. $R_H\Phi$ is in $\mathcal{S}^1(H)$, and (8-1) holds.

(ii) When H is a closed subgroup of G, with Haar measure $d\xi$, then the mapping

$$T_H: \quad [T_H\Phi](\dot{x}) := \int_H \Phi(x+\xi)d\xi, \quad \Phi\in\mathcal{S}^1(G), \quad \dot{x} := \pi(x) \quad [\pi: G \to G/H],$$

is a Banach algebra morphism of $\mathcal{S}^1(G)$ onto $\mathcal{S}^1(G/H)$. Let V* be a compact nd. of 0 in G* and put $W_* := [(-V^*)+V^*]\cap H_*$; if the Haar measures on G, H, G/H are coherent, then we have

$$\|T_H\Phi\|_{W_*} \le \|\Phi\|_{V^*}, \quad \Phi\in\mathcal{S}^1(G). \tag{8-2}$$

Moreover, we also have

$$\|T_H\Phi\|_\infty \le C(H_{V^*})\cdot\|\Phi\|_{V^*}, \quad \Phi\in\mathcal{S}^1(G), \tag{8-3}$$

the constant $C_H(V^*)$ being as in § 1.2 (cf. Feichtinger [7, Th.7, B, p.281).

The restriction inequality for $\mathcal{S}^1(G)$ [(5-1)], applied to Φ_a [$\Phi_a(x) := \Phi(x+a)$, $a\in G$], shows that $T_H\Phi$ is defined everywhere on G/H and satisfies (8-3); applying this to $L_y\Phi - \Phi$, we obtain the (uniform) continuity of $T_H\Phi$. That $T_H\Phi$ is in $L^1(G/H)$ and satisfies $\|T_H\Phi\|_1 \le \|\Phi\|_1$ (with the Haar measures as stated) is a familiar fact [§ 1, (1-2)]; also, if $T_H\mathcal{S}^1(G)$ is provided with the quotient norm, then it is a Segal algebra in $L^1(G/H)$, by a general result [16, Ch.6, § 2.7,

p. 131], and it is easily seen to be invariant. To show that $T_H\mathfrak{S}^1(G)$ is in $\mathfrak{S}^1(G/H$, and that (8-2) holds, let (*) $\Phi = \sum_{n\geq 1} \Phi_n$ be a V^*-representation of $\Phi \in \mathfrak{S}^1(G)$; from (*) we shall derive a W_*-representation of $T_H\Phi$, with W_* as stated. For this, note that (**) $T_H\Phi = \sum_{n\geq 1} T_H\Phi_n$, where $\sum_{n\geq 1} \|T_H\Phi_n\|_1 < \infty$, [cf. § 1 (1-2)], and $\sum_{n\geq 1} \|T_H\Phi_n\|_\infty < \infty$ by (8-3). Now for $T_H\Phi$ only those Φ_n in (*) matter for which $T_H\Phi_n \neq 0$, i.e. the function defined by the series containing only those Φ_n is in $\mathfrak{S}^1(G)$ and has the same T_H-image as Φ; we may thus assume $T_H\Phi_n \neq 0$, $n \geq 1$. Since $\mathcal{F}_{G/H}T_H\Phi_n = R_{H_*}\hat{\Phi}_n$, the assumption entails $H_* \cap \operatorname{Supp} \hat{\Phi}_n \neq \emptyset$, $n \geq 1$. Here $\operatorname{Supp} \hat{\Phi}_n \subset a_n^*+V^*$, for some $a_n^* \in G^*$; but if for an $a^* \in G^*$ we have $H_*\cap(a^*+V^*) \neq \emptyset$, i.e. $a^*+v^* = a_*$ for some $a_* \in H_*$, $v^* \in V^*$, then we have $a^*+V^* \subset a_*+W^*$ with $W^* := (-V^*)+V^*$. a nd. of 0 in G^*, and since a_* is in H_*, we clearly have $H_*\cap(a_*+W^*) = a_*+W_*$, with $W_* := H_*\cap W^*$, a nd. of 0 in H_*. Thus $T_H\Phi_n$ is in $\mathfrak{S}^{W_*}(G/H)$ for each $n \geq 1$ and (**) is a W_*-representation; hence $T_H\Phi$ is in $\mathfrak{S}^1(G/H)$. The inequality (8-2) follows at once.

We have given separate proofs for R_H and T_H; but one can also deduce either of these results from the other by using the Fourier transformation [§ 2.6 (iii)] and one of the relations

$$T_H = \mathcal{F}_{G/H}^{-1} R_{H_*}\mathcal{F}_G , \quad R_H = \mathcal{F}_H^{-1} T_{H_*}\mathcal{F}_G$$

(cf. Feichtinger [7, p. 282]). The proofs above yield the additional information (8-1) and (8-2) about the norms; cf. in this respect the final remarks in § 2.6 (iii).

(iii) Let K be a <u>compact</u> subgroup of G; denote by $\mathfrak{S}^1(G;K)$ the Banach subalgebra of $\mathfrak{S}^1(G)$ consisting of all K-periodic functions in $\mathfrak{S}^1(G)$. <u>The mapping</u> $\dot{f} \longmapsto \dot{f}\circ\pi_K$, $\dot{f} \in \mathfrak{S}^1(G/K)$ [π_K: $G \longrightarrow G/K$], <u>is a Banach algebra isomorphism of</u> $\mathfrak{S}^1(G/K)$ <u>onto</u> $\mathfrak{S}^1(G;K)$; moreover, if V^* is a compact nd. of 0 in the open subgroup K_* of G^* (and hence in G^*), then we have, if $m_K(K) = 1$,

$$\|\dot{f}\circ\pi_K\|_{V^*} = \|\dot{f}\|_{V^*} , \quad \dot{f} \in \mathfrak{S}^1(G/K). \tag{8-4}$$

Let, for $\dot{f} \in \mathfrak{S}^1(G/K)$, $\dot{f} = \sum_{n\geq 1} \dot{f}_n$ be a V^*-representation in $\mathfrak{S}^1(G/K)$. Then $\dot{f}\circ\pi_K = \sum_{n\geq 1} \dot{f}_n\circ\pi_K$ is a V^*-representation in $\mathfrak{S}^1(G)$: $\mathcal{F}_G[\dot{f}\circ\pi_K]$ is simply the zero-extension of $\mathcal{F}_{G/H}\dot{f}_n$ from K_* to G^*, and $\|\dot{f}\circ\pi_K\|_1 = \|\dot{f}\|_1$; thus $\|\dot{f}\circ\pi_K\|_{V^*} \leq \|\dot{f}\|_{V^*}$. Conversely, if $\dot{f}\circ\pi_K \in \mathfrak{S}^1(G;K)$ and $\dot{f}\circ\pi_K = \sum_{n\geq 1} f_n$ is a V^*-representation in $\mathfrak{S}^1(G)$, then $\dot{f} = \sum_{n\geq 1} T_H f_n$ is a V^*-representation in $\mathcal{S}^1(G/K)$: $\|T_H f_n\|_1 \le \|f_n\|_1$ and $\mathcal{F}_{G/H}[T_H f_n]$ is the restriction of $\mathcal{F}_G f_n$ to K_*. Hence $\|\dot f\|_{V*} \le \|\dot f \circ \pi_K\|_{V*}$, and (8-4) follows.

(iv) Let H be an <u>open</u> subgroup of G. For a function f on H we define the 'zero extension' $E_H f$ of f to G by $[E_H f](x) := f(x)$ for $x \in H$, $[E_H f](x) := 0$ for $x \in G$, $x \notin H$. Let $\mathcal{S}^1(G)_H$ be the Banach subalgebra of $\mathcal{S}^1(G)$ consisting of all $f \in \mathcal{S}^1(G)$ which have support in H. <u>The mapping</u> $f \longmapsto E_H f$, $f \in \mathcal{S}^1(H)$, <u>is a Banach algebra isomorphism of</u> $\mathcal{S}^1(H)$ <u>onto</u> $\mathcal{S}^1(G)_H$; moreover, if V* is a compact nd. of 0 in $G^*/H_* \cong H^*$, then

$$\|E_H f\|_{\pi_*^{-1}(V*)} = \|f\|_{V*} \quad [\pi_*: G^* \longrightarrow G^*/H_*].$$

E_H obviously preserves the L^1-norm; moreover, if $f \in \mathcal{S}^1(G)_H$ and $f = \sum_{n\ge1} f_n$ is a V*-representation in $\mathcal{S}^1(H)$, then $E_H f = \sum_{n\ge1} E_H f_n$ is a $\pi_*^{-1}(V*)$-representation in $\mathcal{S}^1(G)$, since $\mathcal{F}_G[E_H f_n] = [\mathcal{F}_H f_n]\circ\pi_*$. Hence $\|E_H f\|_{\pi_*^{-1}(V*)} \le \|f\|_{V*}$. Conversely, let f be a function on H such that $E_H f$ has a $\pi_*^{-1}(V*)$-representation $E_H f = \sum_{n\ge1} g_n$ in $\mathcal{S}^1(G)$, and put $g_n' = E_H R_H g_n$ [R_H: restriction to H]; then $\|g_n'\|_1 \le \|g_n\|_1$, $\mathcal{F}_G g_n' = [T_{H_*}\hat g_n]\circ\pi_*$ [(cf. [16, Ch.5, § 5.5 (ii), p.122]). Thus $E_H f = \sum_{n\ge1} g_n'$ is also a $\pi_*^{-1}(V*)$-representation in $\mathcal{S}^1(G)$; it follows that $f = \sum_{n\ge1} R_H g_n$ is a V*-representation in $\mathcal{S}^1(H)$ and $\|f\|_{V*} \le \|E_H f\|_{\pi^{-1}(V*)}$, whence equality results.

For properties (iii) and (iv) above, cf. Feichtinger [7, Lemma 8, p.283]; these properties (and their proofs) are 'dual' to another.

<u>2.9</u> $\mathcal{S}^1(G)$ has yet another kind of functorial property, based on the following definition. Let G be a l.c.a. group, H an <u>open</u> subgroup; let $S^1(H)$ be a Segal algebra in $L^1(H)$. We define <u>the Segal algebra induced by</u> $S^1(H)$ <u>in</u> $L^1(G)$ as follows: Let Γ be a complete system of representatives of G (mod H), i.e. a subset of G containing exactly one element of each coset of H in G; for $f \in L^1(G)$ we define f_γ, $\gamma \in \Gamma$, by $f_\gamma(x) := f(\gamma + x)$, $x \in H$, so that $f_\gamma \in L^1(H)$. Then we define

$$S^1(H)^G := \left\{ \Phi \mid \Phi \in L^1(G),\ \Phi_\gamma \in S^1(H),\ \gamma \in \Gamma,\ \sum_{\gamma\in\Gamma} \|\Phi_\gamma\|_{S^1(H)} < \infty \right\}$$

and put

$$\|\Phi\| := \sum_{\gamma\in\Gamma} \|\Phi_\gamma\|_{S^1(H)}.$$

It is readily verified that $S^1(H)^G$ is independent of the choice of Γ and, with the norm above, is a Segal algebra in $L^1(G)$; also, if $S^1(H)$ is invariant, then so is $S^1(H)^G$.

$S^1(G)$ and $S^1(H)$ satisfy the relation

$$S^1(G) = S^1(H)^G \quad (H \text{ an open subgroup of } G).$$

We show that $S^1(H)^G \subset S^1(G)$. Let $\Phi \in S^1(H)^G$; let $\Phi_\gamma^{(0)}$ be the function on G equal to Φ on the coset $\gamma+H$, $\gamma \in \Gamma$, and zero outside. Then $\Phi(x) = \sum_{\gamma\in\Gamma} \Phi_\gamma^{(0)}(x)$, $x \in G$, and by § 2.8 (iv) we have (since $\Phi_\gamma^{(0)}$ is a translate of $E_H[\Phi_\gamma]$): $\Phi_\gamma^{(0)} \in S^1(G)$, $\|\Phi_\gamma^{(0)}\|_{S^1(G)} = \|\Phi_\gamma\|_{S^1(H)}$, by appropriate choice of the norms, and hence, by the very definition of $S^1(H)^G$, $\sum_{\gamma\in\Gamma} \|\Phi_\gamma^{(0)}\|_{S^1(G)} < \infty$; thus $\Phi \in S^1(G)$. If we use the norm $\|\ \|_{V*}$ for $S^1(H)$ determined by a compact nd. V* of 0 in H*, then the above norm is simply that determined by $\pi_*^{-1}(V^*) \subset G^*$, as is readily seen [cf. § 2.8 (iv)].

REMARK. If G contains a compact open subgroup, we thus have a particularly simple expression for the functions in $S^1(G)$.

2.10 By using § 2.9 we can describe the structure of $S^1(G)$.

(i) Let $G = H\times\Gamma$, with Γ a discrete group. Then, if we put, for $\Phi = S^1(G)$, $\Phi_\gamma(x) := \Phi(x,\gamma)$, $(x,\gamma) \in H\times\Gamma$, the mapping $\Phi \longrightarrow \underline{\Phi}$, where $\underline{\Phi}(\gamma) = \Phi_\gamma$, is an isomorphism of $S^1(G)$, as a Banach algebra, onto the Banach algebra $L^1(\Gamma, S^1(H))$ (with convolution as multiplication). It is of interest to observe that this can be exhibited as a special case of the tensor representation § 2.6 (ii).

(ii) Let $G = H\times K$, with K a compact group. Then $S^1(G)$ consists of all functions on G such that

$$\Phi(x,k) = \sum_{k^*\in K^*} \Phi_{k^*}(x)\cdot\chi_{k^*}(k) \ , \ \Phi_{k^*} \in S^1(H), \quad (x,k) \in H\times K \ ,$$

where $\sum_{k^*\in K^*} \|\Phi_{k^*}\|_{S^1(H)} < \infty$, and the norm in $S^1(G)$ may be taken as

$$\|\Phi\|_{S^1(G)} := \sum_{k^*\in K^*} \|\Phi_{k^*}\|_{S^1(H)} \ .$$

Thus $S^1(H\times K)$ is, by an obvious mapping, isomorphic, as a Banach algebra, to the algebra (under convolution) of 'absolutely convergent' Fourier series on K with coefficients in $S^1(H)$. This is dual to (i).

Both (i) and (ii) have obvious analogues for arbitrary discrete or compact subgroups of G.

(iii) For a general l.c.a. group G, we use structure theory: G contains an open subgroup $H \cong \mathbb{R}^\nu\times K$, K compact. We can describe $S^1(H)$ in terms of $S^1(\mathbb{R}^\nu)$ by (ii), and $S^1(G)$ in terms of $S^1(H)$ by § 2.9. For elementary groups, or compactly generated groups, of course, (i) and

(ii) suffice.

2.11 It is possible to characterize $\mathfrak{S}^1(G)$ among Segal algebras by functorial properties, as Losert [11, Th. 1, p. 133] has shown. This characterization, in a slightly modified form, is as follows; it is, of course, not needed for Weil's theory.

Suppose that to a general l.c.a. group G a Segal algebra $S^1(G)$ *is associated in such a way that the following conditions hold*:

(0) $S^1(G)$ *consists of continuous functions*.

(i) *If* ρ *is an isomorphism of* G *onto* G', *then* $\Phi \longmapsto |\rho|^{-1}\cdot[\Phi\circ\rho^{-1}]$, $\Phi\in S^1(G)$, *maps* $S^1(G)$ *onto* $S^1(G')$; *in particular*, $S^1(G)$ *is invariant under automorphisms of* G.

(ii) *The Fourier transform maps* $S^1(G)$ *onto* $S^1(G^*)$.

(iii) *For any closed subgroup* H *of* G, *the restriction operator* R_H *maps* $S^1(G)$ *into* $S^1(H)$.

(iv) *If* f, g *are in* $S^1(G)$, *then* $f\otimes g$ *is in* $S^1(G\times G)$.

(v) *If* H *is an open subgroup of* G, *then* $S^1(G)$ *is contained in* $S^1(H)^G$, *the Segal algebra induced in* $L^1(G)$ *by* $S^1(H)$ [§ 2.9].

Then $S^1(G) = \mathfrak{S}^1(G)$.

The role of condition (0) is that it makes (ii) and (iii) simple to state. For condition (i), see § 2.6 (i); it says that $S^1(G)$ is well-defined. For (ii), see § 2.6 (iii); for (iii), (iv), (v) see, respectively, § 2.8 (i), § 2.6 (ii), and § 2.9.

2.12 The proof of the result in § 2.11 is based on the following lemma.

LEMMA. *Let* Γ *be a discrete subgroup of* G *such that* G/Γ *is compact, so that* Γ_* *is also discrete and* G^*/Γ_* *compact. There is a* g *in* $\Lambda^{\mathcal{K}}(G)$ *such that* [cf. § 1.2, Remark, and also § 1.9, Remark]

$$\sum_{\gamma_*\in\Gamma_*} \hat{g}(x^*+\gamma_*) = 1 \quad \text{for all } x^*\in G^*. \tag{12-1}$$

If $f\in L^1(G)$ *is continuous and satisfies*

$$\sum_{\gamma_*\in\Gamma_*} \|f*(\chi_{\gamma_*}\cdot g)\|_1 < \infty , \tag{12-2}$$

then f *is in* $\mathfrak{S}^1(G)$.

The proof is very simple. If (12-1) holds, then we can write $\hat{f} = \sum_{\gamma_*\in\Gamma_*} \hat{f}\cdot[L_{\gamma_*}\hat{g}]$, in the sense of pointwise convergence. Hence, as $\hat{f}\cdot[L_{\gamma_*}\hat{g}]$ is the F.t. of

$$f_{\gamma_*} := f*[\chi_{\gamma_*}\cdot g] ,$$

we shall have, if (12-2) holds,

$$f = \sum_{\gamma_*\in\Gamma_*} f_{\gamma_*} , \quad \sum_{\gamma_*\in\Gamma_*} \|f_{\gamma_*}\|_1 < \infty ,$$

and here each f_{γ_*} is in $\mathfrak{S}^{V*}(G)$, if V* is a compact nd. of 0 in G*, a translate of which covers Supp $\hat{g}$. Hence $f_{\gamma_*} \neq 0$ only for countably many $\gamma_* \in \Gamma_*$, and f lies in $\mathfrak{S}^1(G)$.

REMARK. It can be shown that _Lemma 2.12 extends to arbitrary discrete subgroups of_ G (in particular, $\Gamma = \{0\}$), the sums in (12-1), (12-2) being then replaced by integrals, of course. In the converse direction, we mention that _for arbitrary closed subgroups_ H _of_ G _the following relation holds_:

$$\int_{H_*} \|f*[\chi_{\xi_*}\cdot g]\|_{V*} d\xi_* \le C\cdot \|f\|_{V*}\cdot \|g\|_{V*} \quad \text{for all } f,\ g \text{ in } \mathfrak{S}^1(G),$$

the constant C depending only on H and the choice of V* for the Segal norm. These brief indications must suffice here; we shall not need these results. See in this context Losert [11, Prop.2, p.131].

_2.13_ We can now prove the result in § 2.11. First we note that the condition (ii) of § 2.11 implies, since $S^1(G^*)$ is invariant under translations: if $f \in S^1(G)$, then $\chi_{a*}\cdot f \in S^1(G)$ for all $a^* \in G^*$. We define a norm $\|\ \|'$ in $S^1(G)$ by

$$\|f\|' := \|\hat{f}\|_{S^1(G^*)} , \quad f \in S^1(G),$$

and for this norm $S^1(G)$ is also a Segal algebra [we can apply § 2.2 to $S^1(G^*)$!]. Thus there is a constant C such that for the norm $\|\ \|_S$ in $S^1(G)$ the following relation holds [cf. § 2.1, Remark 3]:

$$C^{-1}\cdot \|f\|' \le \|f\|_S \le C\cdot \|f\|', \quad f \in S^1(G).$$

Replacing here f by $\chi_{a*}\cdot f$, we obtain, since $\|f\|'$ is invariant under this transformation,

$$C^{-2}\cdot \|f\|_S \le \|\chi_{a*}\cdot f\|_S \le C^2\cdot \|f\|_S \quad \text{for all } a^* \in G^*.$$

Hence also

$$C^{-2}\cdot \|f\|_S \le \sup_{a^*\in G^*} \|\chi_{a*}\cdot f\|_S \le C^2\cdot \|f\|_S .$$

Thus

$$\|f\|_S' := \sup_{a^*\in G^*} \|\chi_{a*}\cdot f\|_S , \quad f \in S^1(G),$$

is a norm for which $S^1(G)$ is an _invariant_ Segal algebra. We may now suppose that the Segal algebra $S^1(G)$ is already invariant for the original norm $\| \; \|_S$.

By condition (ii) in § 2.11 again, $S^1(G)$ is contained in the invariant Segal algebra $\Lambda^1(G)$ with the norm (2-1). Hence for some constant C_1 we have [cf. § 2.1, Remark 4]

$$\|f\|_{\Lambda^1} \le C_1 \cdot \|f\|_S \quad \text{for all } f \in S^1(G). \tag{13-1}$$

We can now proceed further as follows. Let Γ be a discrete subgroup of G with G/Γ compact, so that Γ_* is a discrete subgroup of G^* and G^*/Γ_* is compact. Then there is a g in $\Lambda^{\mathcal{K}}(G)$, and hence in $S^1(G)$ [§ 2.1 (A)], satisfying condition (12-1) of the lemma in § 2.12. We will show that any $f \in S^1(G)$ satisfies the condition (12-2) of that lemma; so, by the lemma, $f \in \mathcal{S}^1(G)$. Thus $S^1(G) \subset \mathcal{S}^1(G)$, which yields equality, since $S^1(G)$ has been shown to be invariant.

To prove that any $f \in S^1(G)$ satisfies (12-2), note that $\hat{f}$, $\hat{g}$ are in $S^1(G^*)$ by condition (ii) of § 2.11 and that, by (iv), $\hat{f} \otimes \hat{g} \in S^1(G^*)$. Put

$$\hat{\Phi}(x^*, y^*) = \hat{f}(x^*) \cdot \hat{g}(x^* - y^*) \;, \quad (x^*, y^*) \in G^* \times G^* \,,$$

so by (i) $\hat{\Phi}$ lies in $S^1(G^* \times G^*)$. Consider the subgroup $G^* \times \Gamma_*$ of $G^* \times G^*$: the restriction of $\hat{\Phi}$ to $G^* \times \Gamma_*$ lies in $S^1(G^* \times \Gamma_*)$ by § 2.11, (iii). Now $G^* \times \Gamma_*$ contains the open subgroup $G^* \times \{0\}$, so we can apply § 2.11, (v): putting

$$\hat{\Phi}_{\gamma_*}(x^*) = \hat{\Phi}(x^*, \gamma_*) \;, \; x^* \in G^*, \text{ for } \gamma_* \in \Gamma_* \,,$$

we thus have

$$\sum_{\gamma_* \in \Gamma_*} \|\hat{\Phi}_{\gamma_*}\|_{S^1(G^*)} < \infty \,. \tag{13-2}$$

Here $\hat{\Phi}_{\gamma_*}$ is simply $\hat{f} \cdot L_{\gamma_*} \hat{g}$; applying (13-1) to $S^1(G^*)$, we obtain that

$\sum_{\gamma_* \in \Gamma_*} \|\hat{f} \cdot L_{\gamma_*}\|_{\Lambda^1(G^*)} < \infty$, and hence

$$\sum_{\gamma_* \in \Gamma_*} \|f * [\chi_{\gamma *} \cdot g]\|_1 < \infty \,,$$

which is condition (12-2) of the lemma in § 2.12. Thus, <u>if</u> G <u>contains a discrete subgroup</u> Γ <u>such that</u> G/Γ <u>is compact</u>, <u>then</u> $S^1(G) = \mathcal{S}^1(G)$.

REMARK. We can apply the argument leading to (13-2) to <u>any functions</u> f, g <u>in</u> $\mathcal{S}^1(G)$, assuming only that G/Γ is compact, i.e., that Γ_* is discrete. This gives, in particular, a converse to the lemma in § 2.12; the converse extends to general closed subgroups, as indicated in § 2.12, Remark.

It is now easy to finish the proof. For any l.c.a. group H generated by a compact nd. of 0 we have, $S^1(H) = \mathcal{S}^1(H)$, since such an H contains a discrete subgroup Γ with H/Γ compact (this is the first step in the structure theory of l.c.a. groups); any l.c.a. group obviously contains an <u>open</u> subgroup H of that type, and so we can apply condition (v) of § 2.11: since we already know that $S^1(H) = \mathfrak{S}^1(H)$, we obtain

$$S^1(G) \subset \mathfrak{S}^1(H)^G = \mathfrak{S}^1(G) \quad \text{[cf. § 2.9]}.$$

Hence $S^1(G) = \mathfrak{S}^1(G)$, since $S^1(G)$ is invariant. Thus the result in § 2.11 is completely proved.

It will be observed that the condition (v) of § 2.11 plays an essential part in the proof. There exists, for every l.c.a. group G, a Segal algebra $S^1(G)$ which satisfies conditions (O)-(iv) of § 2.11, but which, as Losert [11, Th.2, p.136] has shown, is distinct from $\mathfrak{S}^1(G)$ unless, of course, G is discrete or compact; this Segal algebra was introduced by R. Bürger [cf. loc. cit. for a reference].

<u>2.14</u> We shall now prove that <u>the space $\mathcal{S}(G)$ of Schwartz-Bruhat functions is contained in</u> $\mathfrak{S}^1(G)$. This was first shown for $G = \mathbb{R}^\nu$ by Poguntke [15, Satz 3] and extended to general l.c.a. G by Feichtinger [7, Th.9]. The proof below, a modification of that of Poguntke, will show that <u>this extends to larger function spaces on</u> G, and that $\mathcal{S}(G)$ <u>is a rather small part of</u> $\mathfrak{S}^1(G)$; for the definition and properties of $\mathcal{S}(G)$ see e.g. [25, n° 11], [9, p.193], [19, § VI]. It will also be shown that $\mathcal{S}(G)$ <u>is continuously embedded in</u> $\mathfrak{S}^1(G)$.

<u>2.15</u> Consider first an elementary group E, given in the form § 1 (10-1). We have defined in § 1.10 a Banach space $\mathcal{B}(E)$ of functions on E. We shall prove:

$$\mathcal{B}(E) \subset \mathfrak{S}^1(E) \tag{15-1}$$

and (once the norm in $\mathfrak{S}^1(E)$ has been fixed) there is a constant C such that

$$\|\Phi\|_{\mathfrak{S}^1(E)} \le C\cdot\|\Phi\|_{\mathcal{B}(E)}, \quad \Phi\in\mathcal{B}(E). \tag{15-2}$$

In § 1.10 we also introduced a decreasing sequence of Banach spaces $\mathcal{B}_n(E)$ contained in $\mathcal{B}(E)$, having $\mathcal{S}(E)$ as intersection, and invariant under automorphisms of E. The above result will, in fact, show that <u>the Banach spaces</u> $\mathcal{B}_n(E)$, <u>and</u> $\mathcal{S}(E)$, <u>are continuously embedded in</u> $\mathfrak{S}^1(E)$. Of course, $\mathcal{S}(E)$ <u>is a rather small part of</u> $\mathcal{B}(E)$, <u>thus of</u> $\mathfrak{S}^1(E)$.

For the proof, let first $E = E_0 := \mathbb{R}^\nu\times(\mathbb{R}/\mathbb{Z})^p$ and apply § 1.9, Remark: with Γ, Γ_* and the function g as stated there, we shall prove that for some constant C

$$\sum_{\gamma_*\in\Gamma_*} \|\Phi*(\chi_{\gamma_*}\cdot g)\|_1 \le C\cdot\|\Phi\|_{\mathcal{B}(E_o)} \text{ for all } \Phi\in\mathcal{B}(E_o). \tag{15-3}$$

The lemma in § 2.12 then implies that $\Phi\in\mathcal{C}^1(E_o)$. We take V* as in § 1.9, Remark, to fix the norm in $\mathcal{C}^1(E_o)$. Relation (15-3) shows that (15-2) holds for $E = E_o$:

$$\|\Phi\|_{\mathcal{C}^1(E_o)} \le C\cdot\|\Phi\|_{\mathcal{B}(E_o)} \text{ for all } \Phi\in\mathcal{B}(E_o). \tag{15-4}$$

To prove (15-3), we note: if the weight function w_o on E_o is as in § 1 (9-1), then $w_o\cdot[\Phi*(\chi_{\gamma_*}\cdot g)]$ is a (continuous) bounded function in $L^1(E_o)$; for by § 1 (9-2) we have for $x\in E_o$

$$w_o(x)\cdot\left|\int_{E_o}\Phi(y)\chi_{\gamma_*}(x-y)g(x-y)dy\right| \le \int_{E_o} w_o(y)\,|\Phi(y)|\cdot w_o(x-y)\,|g(x-y)|\,dy,$$

and here the right-hand side is a bounded (continuous) function in $L^1(E_o)$, since $w_o\cdot\Phi$, $w_o\cdot g$ are bounded functions in $L^1(E_o)$ by the definition of $\mathcal{B}(E_o)$ and of g. Thus $w_o\cdot[\Phi*(\chi_{\gamma_*}\cdot g)]$ is a (continuous) function in $L^1\cap L^2(E_o)$.

We can now apply Schwarz's inequality as follows:

$$\|\Phi*(\chi_{\gamma_*}\cdot g)\|_1 \le \|w_o^{-1}\|_2\cdot\|w_o\cdot[\Phi*(\chi_{\gamma_*}\cdot g)]\|_2\,, \quad \gamma_*\in\Gamma_*\,. \tag{15-5}$$

Further, for any function Φ' in $L^2(E_o)$ such that $w_o\cdot\Phi'$ is also in $L^2(E_o)$, we have, in view of the definition of w_o in § 1.9 and with the notation X^λ and others explained there, if we also introduce the obvious notation $|X^\lambda|$:

$$\|w_o\cdot\Phi'\|_2 = \Big\|\sum_{\lambda\le(1)}|X^\lambda|\Phi'\Big\|_2 \le \sum_{\lambda\le(1)}\|\,|X^\lambda|\Phi'\|_2 = \sum_{\lambda\le(1)}\|X^\lambda\Phi'\|_2\,.$$

Thus

$$\|w_o\cdot[\Phi*(\chi_{\gamma_*}\cdot g)\|_2 \le \sum_{\lambda\le(1)}\|X^\lambda[\Phi*(\chi_{\gamma_*}\cdot g)]\|_2 =: S, \text{ say.}$$

We can next apply Plancherel's theorem and the Fourier relation § 1.9 (A*); this gives

$$S = \sum_{\lambda\le(1)}\|\mathcal{F}\{X^\lambda[\Phi*(\chi_{\gamma_*}\cdot g)]\}\|_2 = \sum_{\lambda\le(1)}\|D^{*\lambda}[\hat\Phi\cdot L_{\gamma_*}\hat g]\|_2\,.$$

Thus (15-5) can be changed into the estimate

$$\|\Phi*(\chi_{\gamma_*}\cdot g)\|_1 \le \|w_o^{-1}\|_2\cdot 2^\nu\cdot\max_{\lambda\le(1)}\|D^{*\lambda}[\hat\Phi\cdot L_{\gamma_*}\hat g]\|_2\,, \quad \gamma_*\in\Gamma_*\,. \tag{15-6}$$

Here $D^{*\lambda}[\hat\Phi\cdot L_{\gamma_*}\hat g]$ is a sum of at most 2^ν terms of the form $(D^{*\lambda'}\hat\Phi)\cdot(D^{*\lambda''}L_{\gamma_*}\hat g)$, with $\lambda', \lambda'' \le (1)$; thus, since Supp $L_{\gamma_*}\hat g$ lies in

γ_*+V^*,

$$\max_{\lambda\le(1)} \|D^{*\lambda}[\hat{\Phi}\cdot L_{\gamma_*}\hat{g}]\|_2 \le \tag{15-7}$$
$$\le 2^{\nu}\cdot\sqrt{\{m_{E_o^*}(V^*)\}}\cdot \max_{\lambda\le(1)}\|D^{*\lambda}\hat{g}\|_\infty\cdot \max_{\lambda\le(1)}\ \max_{x^*\in V^*}|D^{*\lambda}\hat{\Phi}(\gamma_*+x^*)| \ .$$

We can estimate the last factor on the right by introducing the polynomial function P_o on $\mathbb{R}^{\nu+p}$ by

$$P_o(t) := \prod_{j=1}^{\nu+p}(1+t_j^2)\ ,\quad t = (t_j)_{1\le j\le\nu+p} \in \mathbb{R}^{\nu+p}\ .$$

Let P_o^* be its restriction to $E_o^* = \mathbb{R}^{\nu}\times\mathbb{Z}^p$; then we can write

$$|D^{*\lambda}\hat{\Phi}(\gamma_*+x^*)| \le \|P_o^*\cdot D^{*\lambda}\hat{\Phi}\|_\infty\cdot P_o^*(\gamma_*+x^*)^{-1},\quad \lambda\le(1),$$

the first factor on the right being finite by the Fourier relation § 1.9 (B). Hence

$$\max_{x^*\in V^*}|D^{*\lambda}\hat{\Phi}(\gamma_*+x^*)| \le \|P_o^*\cdot D^{*\lambda}\hat{\Phi}\|_\infty\cdot 3^{\nu}\cdot P_o^*(\gamma_*)^{-1},\quad \gamma_*\in\Gamma_*,\quad \lambda\le(1), \tag{15-8}$$

since $P_o^*(\gamma_*) < 3^{\nu}\cdot P_o^*(\gamma_*+x^*)$ for $x^*\in V^*$ [note that $1+u^2 < 3\cdot(1+v^2)$ for u, v in $\mathbb{R}$, $|u-v|\le 1$]; P_o^* (or P_o) is a 'function of translation type' [19, § VI, 19, p.36].

Since $\sum_{\gamma_*\in\Gamma_*} P_o^*(\gamma_*)^{-1} < \infty$, we obtain from (15-6), (15-7), (15-8)

$$\sum_{\gamma_*\in\Gamma_*}\|\Phi*(\chi_{\gamma_*}\cdot g)\|_1 \le C_1\cdot\max_{\lambda\le(1)}\|P_o^*\cdot D^{*\lambda}\hat{\Phi}\|_\infty < \infty,\quad \Phi\in\mathcal{B}(E_o), \tag{15-9}$$

were C_1 is a constant, whence $\Phi\in\mathfrak{S}^1(E_o)$, i.e. (15-1) <u>is proved for</u> $E = E_o$.

To obtain (15-3), we apply the Fourier relation § 1.9 (B); this gives

$$P_o^*\cdot D^{*\lambda}\hat{\Phi} = \sum_{\kappa''}\mathcal{F}[D^{\kappa''}X^{\lambda}\Phi],\quad \lambda\le(1),$$

where the summation extends over all $\kappa'' = (\kappa''_j)_{1\le j\le\nu+p}$ such that κ''_j is 0 or 2. Hence

$$\|P_o^*\cdot D^{*\lambda}\hat{\Phi}\|_\infty \le 2^{\nu+p}\cdot\max_{\kappa''}\|D^{\kappa''}X^{\lambda}\Phi\|_1\ ,\ \lambda\le(1),$$

with κ'' as stated. Here $D^{\kappa''}X^{\lambda}\Phi$ is a linear combination of terms $X^{\lambda'}D^{\kappa'}\Phi$ with $\kappa'\le\kappa''$ and $\lambda'\le\lambda$, the coefficients being integers ≥ 1; moreover, the sum of these coefficients is 3^r if there are r ($\ge$0) values of $j\le\nu$ for which $\kappa''_j = 2$ <u>and</u> $\lambda_j = 1$, as induction on ν shows. Thus we have

$$\|D^{\kappa''}X^{\lambda}\Phi\|_1 \le 3^{\nu}\cdot\max\left\{\ \|X^{\lambda'}D^{\kappa'}\Phi\|_1\ \middle|\ \kappa'\le(2),\ \lambda'\le(1)\right\}.$$

Finally, we can write

$$\|X^{\lambda'} D^{\kappa'} \Phi\|_1 \le \|w_o^{-2}\|_1 \cdot \|w_o^3 \cdot D^{\kappa'} \Phi\|_\infty \ , \quad \kappa' \le (2), \ \lambda' \le (1).$$

The last three inequalities show that for some constant C_2

$$\max_{\lambda \le (1)} \|P_o^* \cdot D^{*\lambda} \hat{\Phi}\|_\infty \le C_2 \cdot \max_{\kappa' \le (2)} \|w_o^3 \cdot D^{\kappa'} \Phi\|_\infty \ , \quad \Phi \in \mathcal{B}(E_o).$$

If we combine this with (15-9), we obtain the relation (15-3), with $C = C_1 \cdot C_2$, hence (15-4), i.e. <u>we obtain</u> (15-2) <u>for</u> $E = E_o$.

We have now shown that the relations (15-1) and (15-2) hold for $E = E_o$. Then we have, for general E: the relation (15-2) for E_o implies, in view of the definition of $\mathcal{B}(E)$ and its norm [§ 1.10] and by the property § 2.10 (i) of $\mathfrak{S}^1(E)$, that (15-1) <u>and</u> (15-2) <u>also hold for general elementary groups</u> E. So <u>the results stated at the beginning of</u> § 2.15 <u>are proved</u>.

Hence <u>the results announced in</u> § 2.14 <u>hold for</u> G = E. Then <u>these results follow immediately for general</u> <u>l.c.a. groups</u> G, by the definition of $\mathcal{S}(G)$ and the properties § 2.8 (iii), (iv) of $\mathfrak{S}^1(G)$. Also, from the Banach spaces $\mathcal{B}_n(E)$, $n \ge 1$, mentioned in § 2.15, we can obtain, by the same method of definition as for $\mathcal{S}(G)$, a decreasing sequence of function spaces $\mathcal{B}_n(G)$, each an inductive limit of Banach spaces, which are continuously embedded in $\mathfrak{S}^1(G)$ and have $\mathcal{S}(G)$ as their intersection.

§3. *Weil's unitary operators and the Segal algebra* $\mathfrak{S}^1(G)$

We study here certain automorphisms of $L^2(G)$ introduced by Weil [25, p. 160] and show that they have two fundamental properties: (i) they leave $\mathfrak{S}^1(G)$ invariant; (ii) their restrictions to $\mathfrak{S}^1(G)$ are bounded operators on the Banach space $\mathfrak{S}^1(G)$.

3.1 Let us put for $\alpha \in \mathrm{Aut}(G)$

$$[M(\alpha)\Phi](x) := |\alpha|^{1/2} \cdot \Phi(x\alpha), \quad x \in G, \quad \Phi \in L^2(G).$$

We note that for every $\Phi \in L^2(G)$ the mapping $\alpha \longmapsto M(\alpha)\Phi$ of Aut(G) into $L^2(G)$ is continuous, i.e. the injective unitary representation $\alpha \longrightarrow M(\alpha)$ of the multiplicative group Aut(G) in $L^2(G)$ is continuous. This readily follows from the definition of the topology in Aut(G), in the same way as for $M_1(\alpha)$ in § 1.7 (cf. also the references given there).

3.2 We now introduce the important concept of character of the second degree (Weil [25, n° 1, p. 145]): this is defined as a continuous function ψ: $G \longrightarrow T$ such that

$$\psi(x+y) = \psi(x) \cdot \psi(y) \cdot B(x,y), \quad x, y \text{ in } G,$$

where $B = B_\psi$ is a bicharacter of $G \times G$; thus [cf. § 1.4]

$$B(x,y) = \langle x, y\beta \rangle \quad \text{for some } \beta \in \mathrm{Mor}(G,G^*) \text{ such that } \beta^* = \beta, \tag{2-1}$$

since obviously $B(y,x) = B(x,y)$. We say that β is associated with ψ, and conversely. We now define an automorphism $A(\psi)$ of $L^2(G)$ by (Weil [25, n° 13, p. 160])

$$A(\psi)\Phi := \psi \cdot \Phi, \quad \Phi \in L^2(G).$$

The characters of the second degree obviously form an abelian group under multiplication which we denote by $Ch_2(G)$; $\psi \longmapsto A(\psi)$ is an injective unitary representation of this group. The significance of $Ch_2(G)$ will appear later [§ 4.1].

The symmetric morphism β in (2-1) is uniquely determined by ψ, i.e. $\beta = \beta_\psi$. If ψ_1 and ψ_2 determine the same β, then $\psi_2 = \psi_1 \cdot \chi$ for some character χ of G, and conversely. If $x \longmapsto 2x$ is an automorphism of G, then

$$\psi(x) := \langle x, 2^{-1}x\beta \rangle, \quad \text{with } \beta \in \mathrm{Mor}(G,G^*), \ \beta^* = \beta,$$

is a character of the second degree on G such that $\beta_\psi = \beta$ [note here that $(2^{-1})\beta$ is also symmetric, as is readily seen]. Hence in this case $Ch_2(G)$ is isomorphic, algebraically, to the product of the additive group of all symmetric morphisms of G into G* and the group G*.

REMARK. Even if $x \longmapsto 2x$ is not an automorphism of G, we still have: _given any symmetric_ $\beta \in Mor(G,G^*)$, _there is a_ $\psi \in Ch_2(G)$ _such that_ $\beta_\psi = \beta$. This has been shown by Igusa [9, Lemma 6, p. 202]; the following simple proof was given by M. Burger in his unpublished paper [5] and is included here by his kind permission. We can make $H := G\times T$ into a locally compact group by putting

$$(x_1,t_1)\cdot(x_2,t_2) := (x_1+x_2, t_1t_2\langle x_1, x_2\beta\rangle^{-1}).$$

Since β is symmetric, this group is abelian. Let $T_0 := \{(0,t)\,|\,t\in T\}$, a compact subgroup of H. Put $\chi_0(0,t) := t$ on T_0. Then χ_0 can be extended to a character χ on H, by the duality theorem. Let $\psi(x) := \chi(x,1)$, for $x\in G$. From the relation $(x+y,1) = (x,1)\cdot(y,1)\cdot(0,\langle x,y\beta\rangle)$ we obtain $\psi(x+y) = \psi(x)\cdot\psi(y)\cdot\langle x,y\beta\rangle$.

_3.3_ Suppose that there exists an _isomorphism_ β of G onto G* (not necessarily symmetric). Let a Haar measure on G be given, $|\beta|$ the Haar modulus of β [§ 1.5]. Then we put $\hat{\Phi}^\beta := |\beta|^{1/2}\cdot\hat{\Phi}\circ\beta$, $\Phi\in L^2(G)$; so

$$\hat{\Phi}^\beta(x) := |\beta|^{1/2}\int_G \Phi(y)\langle y,-x\beta\rangle dy, \quad x\in G, \text{ if } \Phi\in L^1\cap L^2(G).$$

We have $\|\hat{\Phi}^\beta\|_2 = \|\Phi\|_2$, and $\Phi \longmapsto \Phi^\beta$ is a unitary operator, or automorphism, of $L^2(G)$; it is, in fact, independent of the choice of the Haar measure. For the applications, it is convenient to write

$$\beta = \gamma^{*-1}, \text{ with } \gamma\in Is(G^*,G).$$

Thus, if $Is(G^*,G) \neq \emptyset$, we can introduce the unitary operators W^γ, $\gamma\in Is(G^*,G)$, in $L^2(G)$ by (cf. Weil [25, n° 13, p. 160])

$$W^\gamma\Phi := |\gamma|^{-1/2}\cdot\hat{\Phi}\circ\gamma^{*-1}, \quad \Phi\in L^2(G). \tag{3-1}$$

In particular,

$$[W^\gamma\Phi](x) = |\gamma|^{-1/2}\int_G \Phi(y)\langle y,-x\gamma^{*-1}\rangle dy, \quad x\in G, \text{ if } \Phi\in L^1\cap L^2(G).$$

The inversion theorem implies that

$$(W^\gamma)^2 = -I \ (I := \text{identity operator on } L^2(G)),\ (W^\gamma)^4 = I.$$

We also note that [cf. § 1.4 (vi)]

$$W^{\gamma\alpha} = W^\gamma M(\alpha), \quad \alpha\in Aut(G), \tag{3-2}$$

with the operator $M(\alpha)$ as above [§ 3.1]. We have written γ as an upper index; as (3-2) shows, we can obtain from a fixed $\gamma \in Is(G^*,G)$ the general operator of this type.

If $\Phi \in L^1 \cap L^2(G)$, we can also express (3-1) in the form

$$[W^{\gamma}\Phi](x) = |\gamma|^{1/2}\int_{G^*} \Phi(x^*\gamma)\langle x,-x^*\rangle dx^* \quad (dx^* \text{ dual to } dx). \qquad (3\text{-}3)$$

<u>3.4 The Weil operators</u> $M(\alpha)$, $A(\psi)$, W^{γ} <u>induce on</u> $\mathfrak{S}^1(G)$ <u>automorphisms of</u> $\mathfrak{S}^1(G)$ as a Banach space, i.e. their restrictions to $\mathfrak{S}^1(G)$ are continuous (= bounded) linear bijective transformations $\mathfrak{S}^1(G) \longrightarrow \mathfrak{S}^1(G)$; <u>in particular, they leave</u> $\mathfrak{S}^1(G)$ <u>invariant</u>. These remarkable properties were discovered by Feichtinger [7, Th. 1, p. 273, and Cor. 13, p. 287]; they form the basis for the use of $\mathfrak{S}^1(G)$ in the context of metaplectic groups.

For $M(\alpha)$ this follows at once from § 2.6 (i); for W^{γ} it follows from § 2.6 (iii) and (i). Now consider $A(\psi)$: we introduce $A\mathfrak{S}^1(G) := \{\psi\cdot\Phi \mid \Phi \in \mathfrak{S}^1(G)\}$ and put $\|\psi\cdot\Phi\|_{A\mathfrak{S}} := \|\Phi\|_{\mathfrak{S}}$; let us show that $A\mathfrak{S}^1(G)$ is an invariant Segal algebra. Indeed, for $\beta = \beta_{\psi}$ as in (2-1) we have $\psi(x-y) = \psi(x)\cdot\psi(-y)\cdot\langle x,-y\beta\rangle$ and thus

$$L_y(\psi\cdot\Phi) = L_y\psi\cdot L_y\Phi = \psi\cdot\psi(-y)\cdot\chi_{-y\beta}\cdot L_y\Phi,$$

whence $L_y(\psi\cdot\Phi)$ is also in $A\mathfrak{S}^1(G)$ and

$$\|L_y(\psi\cdot\Phi)\|_{A\mathfrak{S}} = \|\Phi\|_{\mathfrak{S}} = \|\psi\cdot\Phi\|_{A\mathfrak{S}}.$$

Likewise $\chi_{a^*}\cdot(\psi\cdot\Phi) = \psi\cdot(\chi_{a^*}\cdot\Phi)$ is in $A\mathfrak{S}^1(G)$ and $\|\chi_{a^*}\cdot(\psi\cdot\Phi)\|_{A\mathfrak{S}} = \|\Phi\|_{\mathfrak{S}} = \|\psi\cdot\Phi\|_{A\mathfrak{S}}$. Moreover, we also have

$$\|L_y(\psi\cdot\Phi) - \psi\cdot\Phi\|_{A\mathfrak{S}} = \|\psi(-y)\chi_{-y\beta}L_y\Phi - \Phi\|_{\mathfrak{S}} \longrightarrow 0 \quad (y \longrightarrow 0),$$

as is readily seen by the properties of the norm of an invariant Segal algebra. It results that $\mathfrak{S}^1(G) \subset A\mathfrak{S}^1(G)$; replacing $A(\psi)$ by $A(\psi)^{-1} = A(\bar{\psi})$, we obtain the reverse inclusion, i.e. equality. The argument shows that $\mathfrak{S}^1(G)$ is a Segal algebra under the two norms $\Phi \longmapsto \|\Phi\|_{\mathfrak{S}}$, $\Phi \longmapsto \|\psi\cdot\Phi\|_{\mathfrak{S}}$. It follows from § 2.1, Remark 3, that <u>the transformation</u> $\Phi \longmapsto \psi\cdot\Phi$ is bounded; as it is obviously bijective, it <u>is an automorphism of</u> $\mathfrak{S}^1(G)$ as a Banach space.

We have more generally: (i) $M(\alpha)$ transforms any Segal algebra into a Segal algebra, and any invariant Segal algebra into an invariant Segal algebra; (ii) $A(\psi)$ transforms any invariant Segal algebra into an invariant Segal algebra; (iii) W^{γ} transforms any invariant Segal algebra contained in $A^1(G)$ [cf. § 2 (2-1)] into a Segal algebra with the same property.

The automorphisms of $L^2(G)$ which induce automorphisms of $\mathfrak{S}^1(G)$ form a group (the inverse is also continuous, by a classical theorem).

Let us now consider the following general situation. Let G_j, $j = 1, 2$, be l.c.a. groups. Then $L^2(G_1 \times G_2)$ is the Hilbert tensor product of $L^2(G_1)$ and $L^2(G_2)$ and, given $S_j \in \mathrm{Aut}(L^2(G_j))$, we may form the tensor product $S_1 \otimes S_2 \in \mathrm{Aut}(L^2(G_1 \times G_2))$. _If for_ $j = 1,2$ $S_j \in \mathrm{Aut}(L^2(G_j))$ _induces an automorphism of_ $\mathcal{S}^1(G_j)$, _then_ $S_1 \otimes S_2$ _induces an automorphism of_ $\mathcal{S}^1(G_1 \times G_2)$ _and the operator norm_ $|S_1 \otimes S_2|$ _for_ $\mathcal{S}^1(G_1 \times G_2)$ _satisfies_ $|S_1 \otimes S_2| \le |S_1| \cdot |S_2|$, where $|S_j|$ is the operator norm of S_j on $\mathcal{S}^1(G_j)$.

This readily follows from the tensor representation of $\mathcal{S}^1(G_1 \times G_2)$ [§ 2.6 (ii)]. For, consider $\Phi \in \mathcal{S}^1(G_1 \times G_2)$, a representation § 2 (6-1), and let $\Phi_N = \sum_{n=1}^{N} \Phi_{1n} \otimes \Phi_{2n}$; let $S := S_1 \otimes S_2$. By definition, $S\Phi_N = \sum_{n=1}^{N} S_1\Phi_{1n} \otimes S_2\Phi_{2n}$; now the series § 2 (6-1) converges in $L^2(G_1 \times G_2)$, since $\sum_{n\ge1} \|\Phi_{1n}\|_2 \cdot \|\Phi_{2n}\|_2 < \infty$ by § 2 (3-6), and hence we have: (*) $S\Phi = \sum_{n\ge1} S_1\Phi_{1n} \otimes S_2\Phi_{2n}$ in $L^2(G_1 \times G_2)$. On the other hand the assumption implies that each term $S_1\Phi_{1n} \otimes S_2\Phi_{2n}$ is in $\mathcal{S}^1(G_1 \times G_2)$ and that, in the notation of § 2.6 (ii), $\|S_j\Phi_{jn}\|_{(j)} \le |S_j| \cdot \|\Phi_{jn}\|_{(j)}$; hence we have: $\sum_{n\ge1} \|S_1\Phi_{1n}\|_{(1)} \cdot \|S_2\Phi_{2n}\|_{(2)} < \infty$. Thus (*) is a tensor representation of $S\Phi$ in $\mathcal{S}^1(G_1 \times G_2)$. The assertion follows.

_3.5_ We now prove the following property of the Weil operator $M(\alpha)$.

LEMMA. _Let_ V^* _be a compact neighbourhood of_ 0 _in_ G. _There is a neighbourhood_ $\mathcal{U}$ _of_ e _in_ Aut(G) _and a constant_ C_1 _such that_

$$\|M(\alpha)\Phi\|_{V^*} \le C_1 \cdot \|\Phi\|_{V^*} \quad \text{for all } \Phi \in \mathcal{S}^1(G), \text{ if } \alpha \in \mathcal{U}. \tag{5-1}$$

For the proof consider first $M_1(\alpha)$ [§ 1.7] in place of $M(\alpha)$. We know that $M(\alpha)\Phi$, or $M_1(\alpha)\Phi$, also lies in $\mathcal{S}^1(G)$ [§ 2.6 (i)]; let us look at $M_1(\alpha)\Phi$ more closely. Let $\Phi = \sum_{n\ge1} \Phi_n$ be a V^*-representation. Choose another compact nd. V_1^* of 0 in G^* and put $W^* := V^* + V_1^*$, which is again a compact nd. of 0 in G^*. Let $\mathcal{U}_1$ be a nd. of e in Aut(G) such that $\alpha^* \in \mathcal{U}'(V^*, V_1^*)$, a nd. of 0 in Aut($G^*$), for all $\alpha \in \mathcal{U}_1$. Then

$$M_1(\alpha)\Phi = \sum_{n\ge1} M_1(\alpha)\Phi_n, \quad \alpha \in \mathcal{U}_1,$$

is a W^*-representation: see § 1.7 and note that $\|M_1(\alpha)\Phi_n\|_1 = \|\Phi_n\|_1$. It follows that

$$\|M_1(\alpha)\Phi\|_{W^*} \le \|\Phi\|_{V^*}.$$

There is a constant C such that [§ 2 (4-2])

$$\|\Phi\|_{V*} \le C\cdot\|\Phi\|_{W*}, \quad \Phi\in\mathfrak{S}^1(G). \tag{5-2}$$

Replacing here Φ by $M_1(\alpha)\Phi$, we obtain from the last two inequalities

$$\|M_1(\alpha)\Phi\|_{V*} \le C\cdot\|\Phi\|_{V*}, \quad \Phi\in\mathfrak{S}^1(G),\ \alpha\in\mathcal{U}_1. \tag{5-3}$$

We can write

$$M(\alpha)\Phi = |\alpha|^{-1/2}\cdot M_1(\alpha)\Phi,$$

and here $|\alpha|$ is a continuous function on Aut(G) which is 1 for $\alpha = e$. Let $\mathcal{U}_2$ be a nd. of e in Aut(G) such that $|\alpha| \ge 1/4$ for $\alpha\in\mathcal{U}_2$, and put $\mathcal{U} = \mathcal{U}_1\cap\mathcal{U}_2$. Then

$$\|M(\alpha)\Phi\|_{V*} \le 2C\cdot\|\Phi\|_{V*} \text{ for all } \Phi\in\mathfrak{S}^1(G) \text{ if } \alpha\in\mathcal{U},$$

and the lemma is proved.

<u>3.6</u> Another basic property of the operator $M(\alpha)$ is that <u>for every</u> $\Phi\in\mathfrak{S}^1(G)$

$$\|M(\alpha)\Phi - \Phi\|_{\mathfrak{S}} \longrightarrow 0 \quad \text{when } \alpha\longrightarrow e \text{ in Aut(G)}. \tag{6-1}$$

Since $M(\alpha_1\alpha_2) = M(\alpha_1)M(\alpha_2)$, this amounts to the following: <u>for each</u> $\Phi\in\mathfrak{S}^1(G)$, <u>the mapping</u> $\alpha \longmapsto M(\alpha)\Phi$ <u>of</u> Aut(G) <u>into</u> $\mathfrak{S}^1(G)$ <u>is continuous</u>.

We shall first prove the corresponding property of the operator $M_1(\alpha)$ [§ 1 (7-1)]:

$$\|M_1(\alpha)\Phi - \Phi\|_{\mathfrak{S}} \longrightarrow 0 \quad \text{when } \alpha\longrightarrow e \text{ in Aut(G)}. \tag{6-2}$$

To obtain (6-2), take V^*, V_1^*, and $W^* := V^*+V_1^*$ as in § 3.5. Let $\Phi = \sum_{n\ge1}\Phi_n$ be a V^*-representation. Then

$$M_1(\alpha)\Phi - \Phi = \sum_{n\ge1}\left[\, M_1(\alpha)\Phi_n - \Phi_n \,\right]$$

will be a W^*-representation, if $\alpha\in\mathcal{U}_1$, where $\mathcal{U}_1$ is chosen as in § 3.5; also

$$\|M_1(\alpha)\Phi - \Phi\|_{W*} \le \sum_{n\ge1}\|M_1(\alpha)\Phi_n - \Phi_n\|_1 \longrightarrow 0 \quad \text{when } \alpha\longrightarrow e \text{ in Aut(G)}, \tag{6-3}$$

by § 1.7 and dominated convergence, since $\|M_1(\alpha)\Phi_n - \Phi_n\|_1 \le 2\cdot\|\Phi_n\|_1$, $n \ge 1$. Now there is a constant C such that [cf. (5-2)]

$$\|M_1(\alpha)\Phi - \Phi\|_{V*} \le C\cdot\|M_1(\alpha)\Phi - \Phi\|_{W*}, \quad \Phi\in\mathfrak{S}^1(G),\ \alpha\in \text{Aut(G)}.$$

Hence (6-3) implies (6-2). To obtain (6-1) from (6-2), we simply show

$$\|M(\alpha)\Phi - M_1(\alpha)\Phi\|_{V*} \longrightarrow 0, \quad \text{when } \alpha\longrightarrow e \text{ in Aut(G)}, \tag{6-4}$$

i.e. $\big||\alpha|^{-1/2} - 1\big|\cdot\|M_1(\alpha)\Phi\|_{V*} \longrightarrow 0$ $(\alpha\longrightarrow e)$; as by (5-3) the second factor is bounded for $\alpha\in\mathcal{U}_1$, (6-4) certainly holds, so (6-1) is proved.

<u>3.7</u> From the properties of $M(\alpha)$ shown in §§ 3.5, 3.6 we shall now deduce a property of <u>products of Weil operators</u> which will be basic in later applications. For that purpose, it will be best to consider the matter in a general setting.

DEFINITION. Let $T(\alpha)$ be a bounded linear operator on $\mathfrak{S}^1(G)$, for each $\alpha \in \mathrm{Aut}(G)$. We call $\alpha \longmapsto T(\alpha)$ an <u>admissible operator function</u> on $\mathrm{Aut}(G)$ for $\mathfrak{S}^1(G)$ if it has the following properties:

(A) For each $\Phi \in \mathfrak{S}^1(G)$ the mapping $\alpha \longmapsto T(\alpha)\Phi$ of $\mathrm{Aut}(G)$ into $\mathfrak{S}^1(G)$ is continuous, i.e. for each $\alpha' \in \mathrm{Aut}(G)$, $T(\alpha)\Phi \longrightarrow T(\alpha')\Phi$ in $\mathfrak{S}^1(G)$ when $\alpha \longrightarrow \alpha'$ in $\mathrm{Aut}(G)$.

(B) The function $\alpha \longrightarrow |T(\alpha)|$ (operator norm of $T(\alpha)$) is locally bounded on $\mathrm{Aut}(G)$, i.e. for each $\alpha' \in \mathrm{Aut}(G)$ there is a nd. $\mathcal{U}(\alpha')$ and a positive constant $C(\alpha')$ such that $|T(\alpha)| \le C(\alpha')$ holds for all $\alpha \in \mathcal{U}(\alpha')$.

The results in §§ 3.5, 3.6 amount to the fact that $\alpha \longrightarrow M(\alpha)$ <u>is an admissible operator function</u> on $\mathrm{Aut}(G)$ for $\mathfrak{S}^1(G)$; this follows from the relation $M(\alpha\cdot\alpha_1) = M(\alpha)M(\alpha_1)$ for α, α_1 in $\mathrm{Aut}(G)$.

We shall now prove the following general continuity lemma:

LEMMA. <u>Let</u> $\alpha \longmapsto T_j(\alpha)$, $j = 1,\cdots,n$, <u>be admissible operator functions on</u> $\mathrm{Aut}(G)$ <u>for</u> $\mathfrak{S}^1(G)$; <u>let</u> Φ <u>be in</u> $\mathfrak{S}^1(G)$. <u>Then</u>

$$(\alpha_1,\cdots,\alpha_n) \longmapsto [\prod_{j=1}^{n} T_j(\alpha_j)]\Phi$$

<u>is a continuous mapping of</u> $[\mathrm{Aut}(G)]^n$ <u>into</u> $\mathfrak{S}^1(G)$.

REMARK. The lemma implies, in particular, that <u>a product of admissible operator functions is again admissible</u>.

EXAMPLES.

(i) For $\psi \in \mathrm{Ch}_2(G)$ we have: $\psi\circ\alpha$, $\alpha \in \mathrm{Aut}(G)$, is again in $\mathrm{Ch}_2(G)$ and

$$A(\psi\circ\alpha) = M(\alpha)A(\psi)M(\alpha^{-1}). \tag{7-1}$$

Here $\alpha \longmapsto M(\alpha^{-1})$ is also admissible, and the constant operator $A(\psi)$ is trivially admissible. So $A(\psi\circ\alpha)$ <u>is an admissible operator function</u>.

(ii) Suppose there exists an isomorphism γ of G^* onto G. Then $\alpha \longmapsto W^{\gamma\alpha}$ <u>is an admissible operator function</u> [cf. (3-2)].

The <u>proof of the continuity lemma</u> is simply by induction. For $n = 1$ there is nothing to prove. Let $n > 1$ and assume that continuity holds for $[\mathrm{Aut}(G)]^{n-1}$. Put

$$\alpha := (\alpha_1, \cdots, \alpha_{n-1}) \in [\mathrm{Aut}(G)]^{n-1}, \quad T(\alpha) := \prod_{j=1}^{n-1} T_j(\alpha_j).$$

Let $\mathcal{U}_j(\alpha'_j)$ be a nd. of α'_j in Aut(G) such that $|T_j(\alpha_j)| \le A_j$, say, for $\alpha_j \in \mathcal{U}_j(\alpha'_j)$, $1 \le j \le n-1$, and put $A := \prod_{j=1}^{n-1} A_j$, so that

$$|T(\alpha)| \le A, \quad \alpha \in \mathcal{U}(\alpha') := \prod_{j=1}^{n-1} \mathcal{U}_j(\alpha'_j), \quad \alpha' := (\alpha'_1, \cdots, \alpha'_{n-1}) \in [\mathrm{Aut}(G)]^{n-1}.$$

Let $\Phi' := T_n(\alpha'_n)\Phi$. Then we have for $\alpha \in \mathcal{U}(\alpha')$ and α_n, α'_n in Aut(G)

$$\|T(\alpha)T_n(\alpha_n)\Phi - T(\alpha')T_n(\alpha'_n)\Phi\|_{\mathfrak{S}} \le$$
$$\le A\cdot\|T_n(\alpha_n)\Phi - T_n(\alpha'_n)\Phi\|_{\mathfrak{S}} + \|T(\alpha)\Phi' - T(\alpha')\Phi'\|_{\mathfrak{S}},$$

which tends to 0 if $\alpha_n \longrightarrow \alpha'_n$ in Aut(G) and $\alpha \longrightarrow \alpha'$ in $[\mathrm{Aut}(G)]^{n-1}$, by the induction hypothesis. Hence continuity in $[\mathrm{Aut}(G)]^n$ follows. It will be noted that in the proof the boundedness condition (B) above plays a role.

<u>3.8</u> We apply the continuity lemma to the Weil operators discussed above to obtain an auxiliary result which has a significance that will appear in connexion with Weil's metaplectic group [§ 8].

LEMMA 1 (Special continuity lemma). <u>Let</u> $\gamma \in \mathrm{Is}(G^*, G)$, $\psi_0 \in \mathrm{Ch}_2(G)$. <u>The mapping</u>

$$(\alpha_1, \alpha_2, \alpha_3) \longmapsto A(\psi_0 \circ \alpha_1) W^{\gamma} M(\alpha_2) A(\psi_0 \circ \alpha_3)\Phi, \quad \Phi \in \mathfrak{S}^1(G),$$

<u>of</u> $[\mathrm{Aut}(G)]^3$ <u>into</u> $\mathfrak{S}^1(G)$ <u>is continuous</u>.

For the proof we recall that $\alpha \longmapsto A(\psi_0 \circ \alpha)$ is an admissible operator function on Aut(G) [§ 3.7, Remark, and Example 1], and apply the general continuity lemma above [§ 3.7] again.

Another auxiliary result needed later is the following:

LEMMA 2 (Special boundedness lemma). <u>Given</u> γ <u>and</u> ψ_0 <u>as in Lemma</u> 1, <u>there is a neighbourhood</u> $\mathcal{U}'$ <u>of</u> e <u>in</u> Aut(G), <u>and a constant</u> C', <u>such that for all</u> $\Phi \in \mathfrak{S}^1(G)$

$$\|A(\psi_0 \circ \alpha_1) W^{\gamma} M(\alpha_2) A(\psi_0 \circ \alpha_3)\Phi\|_{\mathfrak{S}} \le C' \cdot \|\Phi\|_{\mathfrak{S}} \quad \text{if } (\alpha_1, \alpha_2, \alpha_3) \in \mathcal{U}' \times \mathcal{U}' \times \mathcal{U}'.$$

This is an obvious consequence of the lemma in § 3.5, in view of the relation (7-1).

§4. *Weil's group of operators and related groups*

4.1 We shall now study the close relationship between the Weil operators and the Heisenberg group $\mathbb{A}(G)$ of § 1.8 (Weil [25, n° 13, p. 160): the Weil operators $M(\alpha)$, $A(\psi_G)$, W^{γ} belong to the normalizer of $\mathbb{A}(G)$ in $\mathrm{Aut}(L^2(G))$, denoted by $\mathbb{B}(G)$; this is Weil's group of operators, a topological group in the strong operator topology. Indeed, let S be any of these operators; then we obtain, by a short calculation, for $t\cdot U(z)\in\mathbb{A}(G)$, where $t\in T$:

$$S^{-1}U(z)S = c(z)\cdot U(z'),\ c(z) = c_S(z)\in T,\ z,z' \text{ in } G\times G^*, \tag{1-1}$$

and for $\alpha\in\mathrm{Aut}(G)$, $\psi_G\in\mathrm{Ch}_2(G)$, $\gamma\in\mathrm{Is}(G^*,G)$, if $z = (u,u^*)$,

$$S = M(\alpha):\quad z' = (u\alpha, u^*\alpha^{*-1}),\ c(z) = 1, \tag{1-2a}$$

$$S = A(\psi_G):\quad z' = (u, u^*+u\beta)\ [\beta = \beta_{\psi_G},\ \S\ 3.2],\ c(z) = \psi_G(u), \tag{1-2b}$$

$$S = W^{\gamma}:\quad z' = (u^*\gamma, -u\gamma^{*-1}),\ c(z) = \langle u^*\gamma, -u\gamma^{*-1}\rangle = \langle u,-u^*\rangle. \tag{1-2c}$$

A formula of type (1-1) will hold for general $S\in\mathbb{B}(G)$. The automorphism $t\cdot U(z)\longmapsto t\cdot S^{-1}U(z)S$ of the topological group $\mathbb{A}(G)$ leaves the operators $t\cdot I$, $t\in T$ [I := identity operator on $L^2(G)$], of $\mathbb{A}(G)$ invariant; these constitute precisely the centre of $\mathbb{A}(G)$. In (1-1), for general $S\in\mathbb{B}(G)$, the functions $z\longmapsto z'$ and $z\longmapsto c(z)$ are continuous (cf. below). Also, by the law of multiplication in $\mathbb{A}(G)$ [cf. § 1.8] the relation

$$c(z_1)U(z_1')\cdot c(z_2)U(z_2') = c(z_1+z_2)F(z_1,z_2)U((z_1+z_2)')$$

holds for z_j, z_j' in $G\times G^*$, or

$$(z_1+z_2)' = z_1'+z_2',$$

$$c(z_1)\cdot c(z_2)\cdot F(z_1',z_2') = c(z_1+z_2)\cdot F(z_1,z_2).$$

Thus we write

$$z' = z\sigma,\ \sigma\in\mathrm{Aut}(G\times G^*),$$

and $\psi(z)$ in place of $c(z)$: we then have the relation

$$\psi(z_1+z_2) = \psi(z_1)\psi(z_2)\cdot F(z_1\sigma, z_2\sigma)F(z_1,z_2)^{-1},\ z_1,z_2 \text{ in } G\times G^*, \tag{1-3}$$

which shows that ψ is in $\mathrm{Ch}_2(G\times G^*)$; this is the fundamental addition formula for ψ.

The formula (1-1) now becomes

$$S^{-1}U(z)S = \psi(z)U(z\sigma),\quad z\in G\times G^*. \tag{1-4}$$

In this way we obtain an automorphism of the group A(G); recall that this is isomorphic to $\mathbb{A}(G)$ as a topological group [§ 1.8], so σ and ψ are continuous, as stated above. The transformations (1-4), multiplied by $t \in T$, leave the centre of $\mathbb{A}(G)$ pointwise fixed, as mentioned.

Now consider <u>the group</u> $B_o(G)$ <u>of all automorphisms of</u> A(G) <u>which leave each element of the centre of</u> A(G) <u>fixed.</u> The same calculations as before show that such an automorphism has the form

$$(z,t) \longmapsto (z\sigma, t\cdot\psi(z)), \quad \sigma \in \mathrm{Aut}(G\times G^*),\ \psi \in \mathrm{Ch}_2(G\times G^*), \tag{1-5}$$

with σ and ψ related by (1-3), and conversely. Let us denote the automorphism (1-5) by (σ,ψ) and write it in exponential notation:

$$(z,t)^{(\sigma,\psi)} = (z\sigma, t\cdot\psi(z)).$$

The law of multiplication in $B_o(G)$ is

$$(\sigma,\psi)\cdot(\sigma',\psi') = (\sigma\sigma', \psi\cdot(\psi'\circ\sigma)) \qquad ([\psi'\circ\sigma](x) := \psi'(x\sigma)). \tag{1-6}$$

REMARK. The group $B_o(G)$ is thus a subgroup of the semi-direct product $\mathrm{Aut}(G\times G^*)\times_s \mathrm{Ch}_2(G\times G^*)$, with the law of multiplication (1-6); it is easy to verify directly that the elements (σ,ψ) of this group which verify (1-3) form a subgroup.

If $S \in \mathbb{B}(G)$ yields, through (1-4), the automorphism $(\sigma,\psi) \in B_o(G)$, then we write

$$\pi_o(S) = (\sigma,\psi), \quad S \in \mathbb{B}(G), \tag{1-7}$$

so that $\pi_o(S_1\cdot S_2) = \pi_o(S_1)\cdot\pi_o(S_2)$, i.e. π_o <u>is an algebraic homomorphism of</u> $\mathbb{B}(G)$ <u>into</u> $B_o(G)$. The formulae (1-2) may now be expressed as follows. We can write for $z \in G\times G^*$, $\sigma \in \mathrm{Aut}(G\times G^*)$, in matrix notation,

$$z = (u,u^*),\ z\sigma = (u\alpha+u^*\gamma, u\beta+u^*\delta) = (u,u^*)\begin{pmatrix} \alpha & \beta \\ \gamma & \delta \end{pmatrix},$$

with $\alpha \in \mathrm{Mor}(G,G)$, $\beta \in \mathrm{Mor}(G,G^*)$, $\gamma \in \mathrm{Mor}(G^*,G)$, $\delta \in \mathrm{Mor}(G^*,G^*)$ [cf. also § 1.4]. To (1-2a,b,c) correspond, respectively, the following elements of $B_o(G)$: first

$$m_o(\alpha) := \left(\begin{pmatrix} \alpha & 0 \\ 0 & \alpha^{*-1} \end{pmatrix}, 1 \right), \quad \alpha \in \mathrm{Aut}(G), \tag{1-8a}$$

the zeros denoting zero mappings in Mor(G,G*) and in Mor(G*,G). Next,

$$a_o(\psi_G) := \left(\begin{pmatrix} 1 & \beta \\ 0 & 1 \end{pmatrix}, \psi_G \otimes 1_{G^*} \right), \quad \psi_G \in \mathrm{Ch}_2(G), \tag{1-8b}$$

where β is the symmetric morphism of G into G* associated with ψ_G [§ 3.2] and 1 denotes the identity mapping for G or G*; 1_{G^*} is the constant function 1 on G*, so $\psi_G \otimes 1_{G^*} \in \mathrm{Ch}_2(G\times G^*)$). Finally,

$$w_o^{\gamma} := \left(\begin{pmatrix} 0 & -\gamma^{*-1} \\ \gamma & 0 \end{pmatrix}, \psi_o^{-1} \right), \quad \gamma \in Is(G^*,G), \quad \psi_o(u,u^*) := \langle u,u^* \rangle, \quad (1\text{-}8c)$$

0 denoting again obvious zero mappings. We can now express (1-2) as

$$\pi_o(M(\alpha)) = m_o(\alpha), \quad \pi_o(A(\psi_G)) = a_o(\psi_G), \quad \pi_o(W^{\gamma}) = w_o^{\gamma} .$$

It will be noted that

$$\alpha \longmapsto m_o(\alpha), \quad \psi_G \longmapsto a_o(\psi_G) \qquad (1\text{-}9)$$

<u>are algebraic isomorphisms of</u> Aut(G), $Ch_2(G)$ <u>into</u> $B_o(G)$, respectively. We mention, in addition, the following easily obtained formula, for later reference:

$$\pi_o(U(a,a^*)) = \left(\begin{pmatrix} 1 & 0 \\ 0 & 1 \end{pmatrix}, \chi_{a^*} \otimes \bar{\chi}_a \right), \quad (a,a^*) \in G \times G^*. \qquad (1\text{-}10)$$

REMARK. As we have seen, characters of the 2nd degree on $G \times G^*$ appear in connexion with the group $B_o(G)$. For the group $\mathbb{B}(G)$, their significance is as follows, as may readily be shown. <u>Let</u> ψ <u>be a continuous function on</u> G <u>such that</u> $\Phi \longmapsto \psi \cdot \Phi$ <u>is an operator on</u> $L^2(G)$ <u>belonging to</u> $\mathbb{B}(G)$. <u>Then</u> ψ <u>is a character of the 2nd degree</u>.

<u>4.2</u> Let us now return to the addition formula (1-3). This relation obviously implies that $F(z_1\sigma, z_2\sigma) \cdot F(z_1,z_2)^{-1}$ is invariant if we interchange z_1, z_2, or

$$F(z_1,z_2) \cdot F(z_2,z_1)^{-1} \text{ is invariant under } z_1 \longmapsto z_1\sigma, \; z_2 \longmapsto z_2\sigma. \qquad (2\text{-}1)$$

The subgroup of $Aut(G \times G^*)$ formed by all σ satisfying (2-1) is called the <u>symplectic group</u> associated to G, and denoted by Sp(G) (Weil [25, n° 3, p. 148]). From (2-1) we obtain the condition for $\sigma = \begin{pmatrix} \alpha & \beta \\ \gamma & \delta \end{pmatrix}$ in $Aut(G \times G^*)$ to lie in Sp(G) [see, in this context, § 4.1 and § 1.4]. Let us first note that the duality of $G \times G^*$ and $G^* \times G$ is realized by

$$\langle z,z^* \rangle_{G \times G^*} = \langle u,v^* \rangle_G \cdot \langle v,u^* \rangle_G, \quad z = (u,u^*) \in G \times G^*, \quad z^* = (v^*,v) \in G^* \times G.$$

Next let η be the isomorphism of $G \times G^*$ onto $G^* \times G$ given by

$$(u,u^*) \longmapsto (-u^*,u), \text{ i.e. } z = z\eta, \text{ where } \eta := \begin{pmatrix} 0 & 1 \\ -1 & 0 \end{pmatrix}$$

(with an obvious notation). Then we can write (2-1), on interchanging z_1, z_2,

$$\langle z_1\sigma, z_2\sigma\eta \rangle_{G \times G^*} = \langle z_1, z_2\eta \rangle_{G \times G^*} \quad \text{for all } z_j \in G \times G^*, \; j = 1,2.$$

This condition is in turn equivalent to

$$\sigma\eta\sigma^* = \eta, \qquad (2\text{-}2)$$

and here $\sigma^* \in Aut(G^* \times G)$ is given by

$$\sigma^* = \begin{pmatrix}\alpha^* & \gamma^*\\ \beta^* & \delta^*\end{pmatrix}.$$

Hence (2-2) gives

$$\sigma = \begin{pmatrix}\alpha & \beta\\ \gamma & \delta\end{pmatrix} \in \mathrm{Sp}(G) \iff \alpha\delta^* - \beta\gamma^* = 1,\ \alpha\beta^* = \beta\alpha^*,\ \gamma\delta^* = \delta\gamma^*. \tag{2-3}$$

Thus $\alpha\beta^* \in \mathrm{Mor}(G,G^*)$ and $\gamma\delta^* \in \mathrm{Mor}(G^*,G)$ are symmetric. Taking inverses in (2-2), we obtain $\sigma^{*-1}\eta^{-1}\sigma^{-1} = \eta^{-1}$ or, equivalently,

$$\sigma^{*-1}\eta'\sigma^{-1} = \eta', \text{ with } \eta' := -\eta^{-1}, \tag{2-4}$$

and here η' corresponds to the isomorphism of $G^*\times G$ onto its dual $G\times G^*$ given by

$$(v^*,v) \longmapsto (-v,v^*),\quad z^* \longmapsto z^*\eta',\quad \eta' := \begin{pmatrix}0 & 1\\ -1 & 0\end{pmatrix}.$$

Now η' plays the same role for $G^*\times G$ as η does for $G\times G^*$; thus (2-4) shows that $\sigma \longmapsto \sigma^{*-1}$ <u>is an isomorphism of</u> $\mathrm{Sp}(G)$ <u>onto</u> $\mathrm{Sp}(G^*)$ [cf. § 1.6, Prop. 1, applied to $\mathrm{Aut}(G\times G^*)$ and $\mathrm{Aut}(G^*\times G)$]. From (2-4) we get

$$\sigma^*\eta'\sigma = \eta',$$

which yields the conditions dual to (2-3):

$$\sigma \in \mathrm{Sp}(G) \iff \alpha^*\delta - \gamma^*\beta = 1,\ \gamma^*\alpha = \alpha^*\gamma,\ \beta^*\delta = \delta^*\beta. \tag{2-5}$$

Hence $\gamma^*\alpha \in \mathrm{Mor}(G^*,G)$ and $\beta^*\delta \in \mathrm{Mor}(G,G^*)$ are also symmetric.

REMARKS.

1. Given any $\sigma \in \mathrm{Sp}(G)$, there is a $\psi \in \mathrm{Ch}_2(G\times G^*)$ for which the relation (1-3) holds; we say that ψ <u>belongs to</u> σ.- This is an application of § 3.2, Remark (but compare also § 4.11 below for a special case). Thus <u>the homomorphism</u> $(\sigma,\psi) \longmapsto \sigma$ of $B_o(G)$ <u>into</u> $\mathrm{Sp}(G)$ <u>is surjective</u> (Igusa [9, Prop.2, p.203]). Any other ψ in $\mathrm{Ch}_2(G)$ belonging to σ must then be of the form $\psi' = \chi'\cdot\psi$, where χ' is a character of $G\times G^*$. Thus the kernel of this homomorphism is the subgroup

$$A_o(G) := \left\{ (\ 1\ ,\ \chi_{a^*}\otimes\chi_a\ )\ \middle|\ (a,a^*) \in G\times G^* \right\}\quad [1 := \text{unit matrix}]$$

of $B_o(G)$ which is isomorphic to $G\times G^*$.

2. For $\sigma \in \mathrm{Sp}(G)$ <u>the Haar modulus</u> $|\sigma|_{G\times G^*}$ <u>is</u> 1. This follows from (2-2) [cf. § 1.5].

<u>4.3</u> We shall now study an important decomposition due to Weil [25, Prop. 1, p. 152].

Suppose that $\mathrm{Is}(G^*,G) \neq \emptyset$ and consider $\begin{pmatrix}\alpha & \beta\\ \gamma & \delta\end{pmatrix}$ in $\mathrm{Aut}(G\times G^*)$ with γ in $\mathrm{Is}(G^*,G)$. Then there are unique 'matrices' $\begin{pmatrix}1 & \beta_1\\ 0 & 1\end{pmatrix}$, $\begin{pmatrix}1 & \beta_2\\ 0 & 1\end{pmatrix}$ in $\mathrm{Aut}(G\times G^*)$ such that the product

$$\begin{pmatrix}1 & \beta_1\\ 0 & 1\end{pmatrix}\begin{pmatrix}\alpha & \beta\\ \gamma & \delta\end{pmatrix}\begin{pmatrix}1 & \beta_2\\ 0 & 1\end{pmatrix} \tag{3-1}$$

is of the form $\begin{pmatrix}0 & \beta'\\ \gamma & 0\end{pmatrix}$: we have $\beta_1 = -\alpha\gamma^{-1}$, $\beta_2 = -\gamma^{-1}\delta$. Moreover, if $\begin{pmatrix}\alpha & \beta\\ \gamma & \delta\end{pmatrix}$ is symplectic, then $\alpha\gamma^{-1}$ and $\gamma^{-1}\delta$ are symmetric [cf. (2-5), (2-3)], thus the first and the third factor in the product above are symplectic. Hence in this case the product, i.e. the matrix $\begin{pmatrix}0 & \beta'\\ \gamma & 0\end{pmatrix}$, is also symplectic which means that $\beta' = -\gamma^{*-1}$. Thus every matrix $\begin{pmatrix}\alpha & \beta\\ \gamma & \delta\end{pmatrix}$ in Sp(G) has a unique decomposition

$$\begin{pmatrix}\alpha & \beta\\ \gamma & \delta\end{pmatrix} = \begin{pmatrix}1 & \alpha\gamma^{-1}\\ 0 & 1\end{pmatrix}\begin{pmatrix}0 & \gamma^{*-1}\\ \gamma & 0\end{pmatrix}\begin{pmatrix}1 & \gamma^{-1}\delta\\ 0 & 1\end{pmatrix}. \tag{3-2}$$

Conversely, any product (3-1) has the same γ, for any β_1, β_2 in Mor(G,G*). Thus (3-2) may be taken as the <u>standard form for symplectic matrices</u> $\begin{pmatrix}\alpha & \beta\\ \gamma & \delta\end{pmatrix}$ <u>with</u> $\gamma \in$ Is(G*,G).

Let us show that the decomposition (3-2) can be 'lifted' to $B_0(G)$. Consider $(\sigma,\psi) \in B_0(G)$, $\sigma = \begin{pmatrix}\alpha & \beta\\ \gamma & \delta\end{pmatrix}$, and write $\gamma = \gamma(\sigma)$; suppose γ is in Is(G*,G). We use the notations introduced in (1-8) and want to find characters of the 2nd degree ψ_1, ψ_2 on G associated to $\alpha\gamma^{-1}$, $\gamma^{-1}\delta$, respectively, such that, in accordance with (3-2), the relation

$$(\sigma,\psi) = a_0(\psi_1)\cdot w_0^{\gamma}\cdot a_0(\psi_2) \quad \text{[cf. (1-8a, b, c)]}$$

holds. Since $a_0(\psi_G)^{-1} = a_0(\psi_G^{-1})$ for $\psi_G \in Ch_2(G)$ [cf. (1-9)], this means

$$a_0(\psi_1^{-1})\cdot(\sigma,\psi)\cdot a_0(\psi_2^{-1}) = w_0^{\gamma}. \tag{3-3}$$

By the law of multiplication in $B_0(G)$, we thus have: if a decomposition (3-3) is possible, ψ_1, ψ_2 must satisfy

$$\psi_1(x)^{-1}\psi(x, -x\alpha\gamma^{-1}+x^*)\psi_2(x^*\gamma)^{-1} = \langle x, -x^*\rangle. \tag{3-4}$$

Then we obtain, on putting $x^* = 0$, respectively $x = 0$:

$$\psi_1(x) = \psi(x, -x\alpha\gamma^{-1}), \quad \psi_2(x^*\gamma) = \psi(0, x^*). \tag{3-5}$$

Moreover, ψ_1, ψ_2 are indeed characters of the 2nd degree associated to $\alpha\gamma^{-1}$, $\gamma^{-1}\delta$, respectively, as an application of the addition formula (1-3) will show. Conversely, if we apply the addition formula to ψ, with $z_1 = (x, -x\alpha\gamma^{-1})$, $z_2 = (0, x^*)$, we obtain

$$\psi(x, -x\alpha\gamma^{-1}+x^*) = \psi(x, -x\alpha\gamma^{-1})\psi(0, x^*)\langle x, -x^*\rangle, \tag{3-6}$$

i.e. (3-4) holds if ψ_1, ψ_2 are defined by (3-5). Summarizing, we have proved the following result:

<u>Weil decomposition</u> (Weil [25, Prop. 1, p. 152]). Let $\Omega_0(G)$ consist of all $(\sigma,\psi) \in B_0(G)$ such that $\gamma(\sigma)$ is in Is(G*,G), where $\gamma(\sigma) := \gamma$ if $\sigma = \begin{pmatrix}\alpha & \beta\\ \gamma & \delta\end{pmatrix}$. <u>Every</u> $(\sigma,\psi) \in \Omega_0(G)$ <u>can be represented uniquely in the form</u>

$$(\sigma,\psi) = a_0(\psi_1)\cdot w_0^{\gamma}\cdot a_0(\psi_2) \quad \text{[cf. (1-8a, b, c)]}, \tag{3-7}$$

with $\gamma = \gamma(\sigma)$, and ψ_1, ψ_2 in $Ch_2(G)$ given by (3-5). Conversely, any such product (σ,ψ) with γ in $Is(G^*,G)$ and ψ_1, ψ_2 in $Ch_2(G)$ is in $\Omega_o(G)$ and $\gamma(\sigma) = \gamma$.

The Weil decomposition has the following application.

4.4 PROPOSITION (Weil [25, n° 13, p. 160]). *If* $S \in \mathbb{B}(G)$ *is of the form*

$$S = A(\psi_1)W^{\gamma}A(\psi_2), \quad \gamma \in Is(G^*,G), \ \psi_1, \ \psi_2 \text{ in } Ch_2(G), \tag{4-1}$$

then γ, ψ_1, ψ_2 *are unique. Conversely, given* $(\sigma,\psi) \in B_o(G)$ *such that* $\gamma(\sigma) \in Is(G^*,G)$, *there is an* S *of the form* (4-1) - *hence unique - such that* $\pi_o(S) = (\sigma,\psi)$ [cf. (1-4), (1-7)]; *moreover, for* $\Phi \in L^1 \cap L^2(G)$ $S\Phi$ *is given by*

$$[S\Phi](x) = |\gamma|^{1/2}\int_{G^*} \Phi(x\alpha + x^*\gamma)\psi(x,x^*)dx^* \text{ for all } x \in G, \tag{4-2}$$

if $\sigma = \begin{pmatrix}\alpha & \beta\\ \gamma & \delta\end{pmatrix}$ [dx^* dual to dx]. *In other words,* $(\sigma,\psi) \longmapsto S\Phi$ *is a lifting of* $\Omega_o(G)$ *to* $\mathbb{B}(G)$.

The first two assertions follow at once from the Weil decomposition, since $\pi_o(S) = a_o(\psi_1)\cdot w_o^{\gamma}\cdot a_o(\psi_2)$. Now, for $\Phi \in L^1 \cap L^2(G)$ we can write, on using the Weil decomposition (3-7), and also § 3 (3-3),

$$S\Phi = \psi_1(x)\cdot|\gamma|^{1/2}\int_{G^*} \Phi(x^*\gamma)\psi_2(x^*\gamma)\langle x,-x^*\rangle dx^*,$$

where ψ_1, ψ_2 are given by (3-5), and this is by (3-6)

$$= |\gamma|^{1/2}\int_{G^*} \Phi(x^*\gamma)\psi(x,-\alpha\gamma^{-1}+x^*)dx^*,$$

which yields formula (4-2) above. This gives *a direct expression of the lifting* (4-1) *for* $(\sigma,\psi) \in \Omega_o(G)$ without using the Weil decomposition explicitly.

4.5 We come now to the *automorphism theorem for the Segal algebra* $\mathcal{S}^1(G)$: *every operator* $S \in \mathbb{B}(G)$ *induces an automorphism* $\Phi \longmapsto S\Phi$ of $\mathcal{S}^1(G)$, *as a Banach space; in particular, it leaves* $\mathcal{S}^1(G)$ *invariant* [cf. § 3.4 for the definition of automorphism].

The proof is essentially the same as that given for $\mathcal{S}(G)$ by Weil [25, n°s 8-12, pp. 153-159]; it requires a somewhat extended analysis. Let us put [cf. § 1 (8-1)]

$$U^{\vee}(z) := U(u,-u^*), \quad z = (u,u^*) \in G\times G^*, \tag{5-1}$$

and introduce the operator

$$U^{\vee}_{\varphi} := \int_{G\times G^*} \varphi(z)U^{\vee}(z)dz, \quad \varphi \in \mathcal{S}^1(G\times G^*), \tag{5-2}$$

i.e.

$$[U_\varphi^v f](x) := \int_G \int_{G^*} f(x+u)\overline{\langle x,u^*\rangle}\varphi(u,u^*)du^*du, \quad f\in \mathcal{K}(G). \tag{5-3}$$

The reason for using $U^v(z)$ in (5-2) rather than $U(z)$ will appear shortly. (5-2) may be considered as an integral in the space of bounded operators on $L^2(G)$.

We can write (5-3) on changing x into y and u into x

$$[U_\varphi^v f](y) = \int_G f(x)\Phi(x,y)dx, \quad f\in \mathcal{K}(G), \tag{5-4}$$

with

$$\Phi(x,y) := \int_{G^*} \varphi(x-y,u^*)\overline{\langle y,u^*\rangle}du^*, \quad \varphi\in \mathcal{G}^1(G\times G^*), \tag{5-5a}$$

i.e.

$$\Phi = [\mathcal{F}_2\varphi]\circ\tau, \tag{5-5b}$$

where $\mathcal{F}_2$ denotes the <u>partial Fourier transform</u> of $\varphi\in\mathcal{G}^1(G\times G^*)$ relative to G^*,

$$[\mathcal{F}_2\varphi](x,y) := \int_{G^*} \varphi(x,x^*)\overline{\langle y,x^*\rangle}dx^*,$$

and τ is the automorphism of $G\times G$ given by

$$\tau:\ (x,y) \longmapsto (x-y,y).$$

We know that $\mathcal{F}_2$ is a bijection of $\mathcal{G}^1(G\times G^*)$ onto $\mathcal{G}^1(G\times G)$ [§ 2.6 (iv)]; moreover, τ induces an automorphism of $\mathcal{G}^1(G\times G)$ [§ 2.6 (i)]. Thus the mapping $\varphi \longmapsto \Phi$ given by (5-5) is a bijection of $\mathcal{G}^1(G\times G^*)$ onto $\mathcal{G}^1(G\times G)$. Its inverse is

$$\varphi(x,x^*) = \int_G \Phi(x+y,y)\langle y,x^*\rangle dy, \tag{5-6a}$$

i.e.

$$\varphi = \mathcal{F}_2^{-1}[\Phi\circ\tau^{-1}], \tag{5-6b}$$

where $\mathcal{F}_2^{-1}$ is the corresponding inverse partial F.t. (which is relative to the second factor in $G\times G$). Let us denote the transformation (5-6) by

$$\varphi = \mathsf{T}\Phi := \mathcal{F}_2^{-1}[\Phi\circ\tau^{-1}], \quad \Phi\in\mathcal{G}^1(G\times G), \tag{5-7}$$

and write correspondingly

$$\Phi = \mathsf{T}^{-1}\varphi := [\mathcal{F}_2\varphi]\circ\tau, \quad \varphi\in\mathcal{G}^1(G\times G^*). \tag{5-8}$$

Then T <u>is a bijection of</u> $\mathcal{G}^1(G\times G)$ <u>onto</u> $\mathcal{G}^1(G\times G^*)$; moreover, T <u>preserves the</u> L^2<u>-norm</u>, $\|\mathsf{T}\Phi\|_2 = \|\Phi\|_2$, which follows from the corresponding property of $\mathcal{F}_2^{-1}$:

$$\int_{G^*} |[\mathcal{F}_2^{-1}\Phi](x,x^*)|^2 dx^* = \int_G |\Phi(x,y)|^2 dy, \quad x \in G,$$

whence

$$\int_G \int_{G^*} |[\mathcal{F}_2^{-1}\Phi](x,x^*)|^2 dx^* dx = \int_G \int_G |\Phi(x,y)|^2 dy dx.$$

Hence $\hat{T}$ <u>can be extended, by continuity, to an isomorphism of</u> $L^2(G\times G)$ <u>onto</u> $L^2(G\times G^*)$, since $\mathcal{S}^1(G\times G)$, $\mathcal{S}^1(G\times G^*)$ are dense in $L^2(G\times G)$, $L^2(G\times G^*)$, respectively [cf. § 2.3, Remark 2].

Now let f, g be in $\mathcal{S}^1(G)$ and consider $\Phi := f\otimes\bar{g} \in \mathcal{S}^1(G\times G)$ [note that $\bar{g} \in \mathcal{S}^1(G)$]. Formula (5-6a) shows that

$$[\hat{T}(f\otimes\bar{g})](z) = \int_G [U(z)f](y)\bar{g}(y)dy, \quad f, g \text{ in } \mathcal{S}^1(G),\ z = (x,x^*),$$

or, with the usual notation for the scalar product in $L^2(G)$,

$$[\hat{T}(f\otimes\bar{g})](z) = (U(z)f|g), \quad z \in G\times G^*. \tag{5-9}$$

<u>Relation</u> (5-9) <u>holds generally</u>, i.e. <u>for</u> f, g <u>in</u> $L^2(G)$. For, let f_n, g_n in $\mathcal{S}^1(G)$ tend in $L^2(G)$ to f, g, respectively $(n \to \infty)$. Then the functions $\hat{T}(f_n\otimes\bar{g}_n)$ tend to the limit $\hat{T}(f\otimes\bar{g})$ in $L^2(G\times G^*)$, when $n \to \infty$. But we also have for each $z \in G\times G^*$: $(U(z)f_n|g_n) \to (U(z)f|g)$ $(n \to \infty)$. Hence $z \longmapsto (U(z)f|g)$, $z \in G\times G^*$, is this limit. Also, $\varphi := \hat{T}(f\otimes\bar{g})$ is a continuous function on $G\times G^*$, for arbitrary f, g in $L^2(G)$.

The relation (5-9) is the reason for the choice of the definition (5-2), (5-3).

We can now prove the automorphism theorem. Let S be in $\mathbb{B}(G)$; define $\bar{S} \in \mathrm{Aut}(L^2(G)$ by

$$\bar{S}g := (S\bar{g})^-, \quad g \in L^2(G), \tag{5-10}$$

so that $\bar{S}\,\bar{g} = \overline{Sg}$ ($S \longmapsto \bar{S}$ is an automorphism of $\mathbb{B}(G)$, but we shall not need this). We then have from (5-9) for f, g in $\mathcal{S}^1(G)$:

$$[\hat{T}(Sf\otimes\overline{Sg})](z) = (U(z)Sf|Sg),$$

$$= (S^{-1}U(z)Sf|g) = \varphi(z\sigma)\psi(z) \quad [(\sigma,\psi) \in B_o(G)],$$

with $\varphi = \hat{T}(f\otimes\bar{g})$ as above. Since $f\otimes\bar{g}$ is in $\mathcal{S}^1(G\times G)$, the function φ is in $\mathcal{S}^1(G\times G^*)$. Thus, if we put $\varphi_S(z) := \varphi(z\sigma)\psi(z)$, $z \in G\times G^*$, then φ_S also lies in $\mathcal{S}^1(G\times G^*)$ [§ 3.4]. Hence there is a function $\Phi \in \mathcal{S}^1(G\times G^*)$ such that $\hat{T}\Phi = \varphi_S$. Then the L^2-isometry of $\hat{T}$ yields that

$$\Phi(x,y) = Sf(x)\cdot\overline{Sg(y)} \text{ almost everywhere on } G\times G. \tag{5-11}$$

The lemma in § 1.11 shows (in the non-trivial case) that we may take Sf, Sg to be continuous, and that (5-11) holds everywhere. Let $g \neq 0$ in $\mathcal{S}^1(G)$ be fixed; it follows from § 2.8 (i) that Sf is in $\mathcal{S}^1(G)$ for all $f \in \mathfrak{S}^1(G)$. Thus $f \longmapsto Sf$ is a linear transformation of $\mathfrak{S}^1(G)$ into itself, obviously bijective. To show that it is bounded, we note: the linear transformation $f \longmapsto f\otimes\bar{g}$ of $\mathfrak{S}^1(G)$ into $\mathfrak{S}^1(G\times G)$ is bounded [cf. § 2.6 (ii)], and so is the transformation $\hat{\mathsf{T}}^{-1}$ of $\mathfrak{S}^1(G\times G)$ onto $\mathfrak{S}^1(G\times G^*)$ [cf. § 2.6 (iv)]; again, the transformation $\varphi \longmapsto \varphi_S$, with φ_S as above, of $\mathfrak{S}^1(G\times G^*)$ onto itself is bounded [§ 3.4], and so is the transformation $\hat{\mathsf{T}}^{-1}$ of $\mathfrak{S}^1(G\times G^*)$ back to $\mathfrak{S}^1(G\times G)$. Finally, the restriction of the functions in $\mathfrak{S}^1(G\times G)$ to $G\times\{0\}$ is a bounded transformation onto $\mathfrak{S}^1(G)$. It follows that $S\in\mathbb{B}(G)$ induces a linear, bounded, bijective mapping $\mathfrak{S}^1(G) \to \mathfrak{S}^1(G)$, i.e. a Banach space automorphism. The automorphism theorem is proved.

<u>4.6</u> <u>The kernel of the mapping</u> π_0 [(1-7)] can easily be obtained by the method used to prove the automorphism theorem in § 4.5. <u>If</u> $S\in\mathbb{B}(G)$ <u>commutes with every</u> $U(z)\in\mathbb{A}(G)$, <u>then</u> $S = c\cdot I$, <u>with</u> $c\in T$ [I := identity operator on $L^2(G)$].

Indeed, formula (5-9) shows, when $S^{-1}U(z)S = U(z)$, $z\in G\times G^*$, that

$$[\hat{\mathsf{T}}(Sf\otimes\overline{Sg})](z) = [\hat{\mathsf{T}}(f\otimes\bar{g})](z) \text{ for all } z\in G\times G^* \ [f,g \text{ in } L^2(G)],$$

whence, in the sense of $L^2(G\times G)$,

$$Sf\otimes\overline{Sg} = f\otimes\bar{g} \text{ for all } f,\ g \text{ in } L^2(G). \tag{6-1}$$

We take continuous functions $f \neq 0$, $g \neq 0$ in $L^2(G)$. By the lemma in § 1.11, Sf, Sg may also be taken continuous. Fix g and choose $a\in G$ such that $Sg(a) \neq 0$. Relation (6-1) then yields $Sf = c\cdot f$, with $c = g(a)/Sg(a)$ in T, since S is unitary. Hence we have $Sf = c\cdot f$ for all $f\in L^2(G)$.

This proof is a variant of the final part of the proof of Theorem 1 in Weil's memoir [25, p. 157].

<u>4.7</u> In Weil's theory a certain space of theta functions plays an essential role; we discuss this space here.

Let Γ be a closed subgroup of G. Denote by $\mathcal{K}(G,\Gamma)$ the linear space of all complex-valued continuous functions θ on $G\times G^*$ which have compact support mod $\Gamma\times\Gamma_*$ and satisfy the functional equation

$$\theta(x+\xi, x^*+\xi_*) = \theta(x,x^*)\langle\xi,x^*\rangle^{-1},\quad (x,x^*)\in G\times G^*,\quad (\xi,\xi_*)\in\Gamma\times\Gamma_*. \tag{7-1a}$$

We can write this, by using the bicharacter F on $(G\times G^*)\times(G\times G^*)$ introduced in § 1 (8-2),

$$\theta(z+\zeta) = \theta(z)\cdot F(\zeta,z)^{-1},\quad z\in G\times G^*,\quad \zeta\in\Gamma\times\Gamma_*. \tag{7-1b}$$

Note that this functional equation implies, in particular,

$$\theta(x,x^*+\xi_*) = \theta(x,x^*), \quad (x,x^*)\in G\times G^*, \quad \xi_*\in\Gamma_*, \tag{7-2}$$

so we may consider θ as a function on $G\times(G^*/\Gamma_*)$. Moreover, the absolute value $|\theta|$ is clearly $\Gamma\times\Gamma_*$-periodic and may thus be considered as a function on

$$Q := G\times G^*/\Gamma\times\Gamma_* = (G/\Gamma)\times(G^*/\Gamma_*). \tag{7-3}$$

REMARK. We assume once for all that coherent Haar measures on G, Γ, G/Γ are given. Then we take the dual Haar measures on G^*, $\Gamma_* = (G/\Gamma)^*$, $G^*/\Gamma_* = \Gamma^*$ which are also coherent [§ 1.2]. On Q we take the product measure.

The space $\mathcal{K}(G,\Gamma)$ contains functions $\neq 0$; this is easily seen as follows. For $\Phi\in\mathcal{K}(G)$ we define a function $\tilde{T}_\Gamma\Phi$ on $G\times G^*$, the <u>theta transform</u> of Φ, also denoted by $\tilde{T}\Phi$ or θ_Φ, by

$$\theta_\Phi(z) := [\tilde{T}_\Gamma\Phi](z) := \int_\Gamma \Phi(x+\xi)\langle\xi,x^*\rangle d\xi = \int_\Gamma [U(z)\Phi](\xi)d\xi, \tag{7-4}$$

$$z = (x,x^*)\in G\times G^*$$

[cf. § 1 (8-1)]. Here θ_Φ is a continuous, bounded function on $G\times G^*$: this follows from the fact that for $\Phi\in\mathcal{K}(G)$

$$\int_\Gamma |\Phi(x+\xi) - \Phi(x_0+\xi)|d\xi < \varepsilon \quad \text{if } x-x_0\in U_\varepsilon, \text{ a nd. of 0 in } G,$$

$$\int_\Gamma |\Phi(x+\xi)|d\xi \le C \quad \text{for all } x\in G$$

(cf. the Lemma in [16, Ch. 3, § 3.2, p. 58]). Now take any function $\dot{k}$ in $\mathcal{K}(Q)$ [cf. (7-3)] and put

$$\theta' := (\dot{k}\circ\pi)\cdot\theta_\Phi \quad [\pi\colon G\times G^* \longrightarrow Q].$$

Then θ' is in $\mathcal{K}(G,\Gamma)$, and we can clearly choose Φ and $\dot{k}$ such that θ' takes the value 1, say, at $(0,0)\in G\times G^*$.

We define, for $1 \le p < \infty$, $\mathcal{F}^p(G,\Gamma)$ as the space of <u>all</u> complex-valued functions on $G\times G^*$ satisfying the functional equation (7-1) and such that

$$N_p(\theta) := \left\{ \int_Q^* |\theta(z)|^p dz\right\}^{1/p} < \infty \quad [\text{upper Haar integral on } Q]. \tag{7-5}$$

$\mathcal{F}^p(G,\Gamma)$ is a <u>complete</u> vector space with the semi-norm N_p (the proof given in Bourbaki's now classical 'Intégration' [3, Ch. IV, § 3, n° 3] still goes through in the present case, as is readily seen). The space $\mathcal{L}^p(G,\Gamma)$ can now be defined, as the closure of $\mathcal{K}(G,\Gamma)$ in $\mathcal{F}^p(G,\Gamma)$, and $L^p(G,\Gamma)$ as the associated Banach space, in perfect analogy with the classical case. For $\theta\in\mathcal{L}^p(G,\Gamma)$ the star in (7-5) may be omitted. By a <u>theta function</u> (relative to the subgroup Γ) will be meant any function in $\mathcal{L}^2(G,\Gamma)$; one writes $\theta\in L^2(G,\Gamma)$ rather than $\theta\in\mathcal{L}^2(G,\Gamma)$, and $\|\theta\|_Q$ for

$N_2(\theta)$. This settles the general definition of theta functions.

Now consider $\Phi \in \mathfrak{S}^1(G)$: the theta transform (7-4) of Φ is again a continuous, bounded function on $G\times G^*$; this follows from the restriction inequality for $\mathfrak{S}^1(G)$ [§ 2.5] which gives

$$\int_\Gamma |\Phi(x+\xi) - \Phi(x_0+\xi)|d\xi \le C_\Gamma(V^*)\cdot \|L_{-x}\Phi - L_{-x_0}\Phi\|_{V^*} < \varepsilon \quad \text{if } x-x_0 \in U_\varepsilon,$$

where U_ε is a nd. of 0 in G, and

$$\int_\Gamma |\Phi(x+\xi)|d\xi \le C_\Gamma(V^*)\cdot \|L_{-x}\Phi\|_{V^*} = C_\Gamma(V^*)\cdot\|\Phi\|_{V^*}, \quad x\in G.$$

4.8 We shall now prove (cf. Weil [25, n°s 17, 18]):

PROPOSITION. If $\Phi \in \mathfrak{S}^1(G)$, then $\theta_\Phi := \mathfrak{T}_\Gamma\Phi$, defined by (7-4), lies in $L^2(G,\Gamma)$, and $\|\theta_\Phi\|_Q = \|\Phi\|_2$. Moreover, the function $\dot{x}^* \mapsto \theta(x,x^*)$, $\dot{x}^* \in G^*/\Gamma_*$, is in $L^1(G^*/\Gamma_*)$ and the theta inversion formula

$$\Phi(x) = \int_{G^*/\Gamma_*} \theta_\Phi(x,x^*)d\dot{x}^*, \quad x\in G, \tag{8-1}$$

holds. The image of $\mathfrak{S}^1(G)$ under $\mathfrak{T}_\Gamma$ is dense in $L^2(G,\Gamma)$ and thus the mapping $\mathfrak{T}_\Gamma$ can be extended to an isomorphism of $L^2(G)$ onto $L^2(G,\Gamma)$, still denoted by $\mathfrak{T}_\Gamma$.

The proof is an application of Poisson's formula [§ 2.5 (ii)] to the function $U(z)\Phi$, which is again in $\mathfrak{S}^1(G)$: this gives for $z = (x,x^*)$ in $G\times G^*$ (the Haar measures being chosen as in § 4.7, Remark)

$$\int_\Gamma \Phi(x+\xi)\langle\xi,x^*\rangle d\xi = \int_{\Gamma_*} \hat{\Phi}(-x^*+\xi_*)\langle x,-x^*+\xi_*\rangle d\xi_*.$$

We replace ξ_* by $-\xi_*$ and put

$$\Phi^*(x^*) := \hat{\Phi}(-x^*) = \int_G \Phi(x)\langle x,x^*\rangle dx. \tag{8-2}$$

Then

$$\theta(x,x^*) = \int_{\Gamma_*} \Phi^*(x^*+\xi_*)\overline{\langle x, x^*+\xi_*\rangle}d\xi_*. \tag{8-3}$$

Thus θ, as a function of $\dot{x}^* \in G^*/\Gamma_*$, lies indeed in $L^1(G^*/\Gamma_*)$. On the other hand we have by the inversion theorem, for each $x\in G$,

$$\Phi(x) = \int_{G^*} \hat{\Phi}(x^*)\langle x,x^*\rangle dx^* = \int_{G^*} \Phi^*(x^*)\overline{\langle x,x^*\rangle}dx^* =$$

$$= \int_{G^*/\Gamma_*}\int_{\Gamma_*} \Phi^*(x^*+\xi_*)\overline{\langle x,x^*+\xi_*\rangle}d\zeta_* d\dot{x}^*,$$

and by (8-3) this yields (8-1).

We note further that the restriction of Φ to Γ is in $\mathfrak{S}^1(\Gamma)$ [§ 2.8 (i)], hence also in $L^2(\Gamma)$. Thus we can apply Plancherel's theorem: this gives, by the definition (7-4),

$$\int_{G^*/\Gamma_*} |\theta_\Phi(x,x^*)|^2 d\dot{x}^* = \int_\Gamma |\Phi(x+\xi)|^2 d\xi,$$

whence

$$\int_{G/\Gamma}\int_{G^*/\Gamma_*} |\theta_\Phi(x,x^*)|^2 d\dot{x}^* d\dot{x} = \int_{G/\Gamma}\int_\Gamma |\Phi(x+\xi)|^2 d\xi d\dot{x},$$

or $\|\theta_\Phi\|_Q = \|\Phi\|_2$, as asserted.

To show that the image of $\mathfrak{S}^1(G)$ under $\tilde{T}_\Gamma$ is dense in $L^2(G,\Gamma)$, we observe, first, that θ_Φ - which is a continuous solution of the functional equation (7-1) such that $\|\theta_\Phi\|_Q < \infty$ - certainly lies in $L^2(G,\Gamma)$. Next consider any $\theta' \in \mathcal{K}(G,\Gamma)$ and define Φ' on G by

$$\Phi'(x) := \int_{G^*/\Gamma_*} \theta'(x,x^*)d\dot{x}^*, \quad x\in G.$$

Then Φ' is continuous which can be seen as follows. θ' satisfies relation (7-2]); thus, putting $\theta''(x,\dot{x}^*) := \theta'(x,x^*)$, we obtain a function θ'' on $G\times(G^*/\Gamma_*)$. This function has support in $G\times\dot{K}^*$, where $\dot{K}^* \subset G^*/\Gamma_*$ is compact. Choose any $a\in G$ and a compact nd. U(a); then θ'' is uniformly continuous on $U(a)\times\dot{K}^*$, hence

$$\int_{G^*/\Gamma_*} |\theta'(x,x^*) - \theta'(a,x^*)|d\dot{x}^* = \int_{G^*/\Gamma_*} |\theta''(x,\dot{x}^*) - \theta''(a,\dot{x}^*)|d\dot{x}^*$$
$$< \varepsilon\cdot m_{G^*/\Gamma_*}(\dot{K}^*) \text{ if } x\in U_\varepsilon(a) \subset U(a).$$

Moreover, we have for $\xi\in\Gamma$

$$\Phi'(x+\xi) = \int_{G^*/\Gamma_*} \theta'(x,x^*)\overline{\langle\xi,x^*\rangle}d\dot{x}^*,$$

whence

$$\int_\Gamma |\varphi'(x+\xi)|^2 d\xi = \int_{G^*/\Gamma_*} |\theta'(x,x^*)|^2 d\dot{x}^*.$$

Thus

$$\int_{G/\Gamma}\int_\Gamma |\Phi'(x+\xi)|^2 d\xi dx = \int_{G/\Gamma}\int_{G^*/\Gamma_*} |\theta'(x,x^*)|^2 d\dot{x}^* d\dot{x},$$

i.e. $\|\Phi'\|_2 = \|\theta'\|_Q$, so that Φ' is in $L^2(G)$.

Let $\theta' \in \mathcal{K}(G,\Gamma)$ be fixed and consider θ_Φ, for any $\Phi\in\mathfrak{S}^1(G)$. Then we can replace θ' by $\theta' - \theta_\Phi$ in the argument just given: in view of the theta inversion formula we obtain

$$\|\Phi' - \Phi\|_2 = \|\theta' - \theta_\Phi\|_Q .$$

Since $\mathfrak{S}^1(G)$ is dense in $L^2(G)$ [§ 2.3, Remark 2], this can be made small by proper choice of Φ; hence $\tilde{T}_\Gamma\mathfrak{S}^1(G)$ is dense in $L^2(G,\Gamma)$. This concludes the proof.

As we have seen, the isomorphism of $L^2(G)$ with $L^2(G,\Gamma)$ is established by means of Poisson's formula. The definition of theta functions and the proof above are a modification of those given by Weil [25, n°s 17, 18]; cf. [20, § II], where a certain difficulty is discussed.

The theta functions may be considered as a relativization, in the sense of [16, Ch.4, § 5], of the transform Φ^* [cf. (8-2)]. For $\Gamma = G$ we have $L^2(G,\Gamma) = L^2(G^*)$, and if $\Gamma = \{0\}$, then $L^2(G,\Gamma) = L^2(G)$.

4.9 We come now to one of the main results in Weil's theory, his Theorem 4. It will be practical to prove first a simple lemma.

LEMMA. *Let Γ be a closed subgroup of G. The subgroup*

$$U(\Gamma\times\Gamma_*) := \{\ U(z)\ |\ z\in \Gamma\times\Gamma_*\ \}$$

of $\mathbb{A}(G)$ is precisely the group of all $c\cdot U(z)$ in $\mathbb{A}(G)$ such that

$$\int_\Gamma c\cdot[U(z)\Phi](\xi)d\xi = \int_\Gamma \Phi(\xi)d\xi \quad \text{for all } \Phi\in \mathcal{S}^1(G). \tag{9-1}$$

Obviously, any $U(z)\in U(\Gamma\times\Gamma_*)$ satisfies (9-1). Conversely, if for some $z = (a,a^*)\in G\times G^*$, $c\in T$, (9-1) holds, then $a\in G$ must be in Γ: for, if a is not in Γ, take a nd. V of 0 in G so small that $[(-a)+V]\cap\Gamma = \emptyset$ and choose $\Phi \geq 0$ in $\mathcal{S}^1(G)$ such that $\Phi(0) > 0$ (clearly, there is such a Φ!); then (9-1) cannot hold for this Φ. So $a = \xi_o$, say, $\xi_o\in\Gamma$, and

$$\int_\Gamma c.\Phi(\xi_o+\xi)<\xi,a^*>d\xi = \int_\Gamma \Phi(\xi)d\xi, \quad \Phi\in \mathcal{S}^1(G),$$

or

$$\int_\Gamma \Phi(\xi_o+\xi)[c\cdot<\xi,a^*> - 1]d\xi = 0 \quad \text{for all } \Phi\in \mathcal{S}^1(G).$$

Since $\mathcal{S}^1(G)|\Gamma = \mathcal{S}^1(\Gamma)$ [§ 2.8 (i)] and $\mathcal{S}^1(\Gamma)$ is dense in $L^1(\Gamma)$, it follows that $c\cdot<\xi,a^*> - 1 = 0$ for all $\xi\in\Gamma$; so $c = 1$ and $a^*\in\Gamma_*$.

We can now state Weil's result [25, Th.4] as follows:

THEOREM. *Let Γ be a closed subgroup of G and let $U(\Gamma\times\Gamma_*)$ be the subgroup of $\mathbb{B}(G)$ defined in the above lemma. Let $\mathbb{B}(G,\Gamma)$ consist of all operators in the normalizer of $U(\Gamma\times\Gamma_*)$ in $\mathbb{B}(G)$ which induce on $U(\Gamma\times\Gamma_*)$* (by the action $U(z) \longmapsto S^{-1}U(z)S$) *an automorphism of Haar modulus* 1. *The Poisson-Weil formula*

$$\int_\Gamma S\Phi(\xi)d\xi = \int_\Gamma \Phi(\xi)d\xi, \quad \Phi\in \mathcal{S}^1(G), \tag{9-2}$$

holds for all $S\in \mathbb{B}(G,\Gamma)$; moreover, *$\mathbb{B}(G,\Gamma)$ is the subgroup of all $S\in \mathbb{B}(G)$ satisfying* (9-2). *$\mathbb{B}(G,\Gamma)$ can be obtained by a lifting - described below* [(9-8)] - *of the subgroup $B_o(G,\Gamma)$ of $B_o(G)$ consisting of all $(\sigma,\psi)\in B_o(G)$ such that σ induces on $\Gamma\times\Gamma_*$ an automorphism of*

<u>Haar</u> <u>modulus</u> 1 <u>and</u> ψ <u>is</u> 1 <u>on</u> $\Gamma\times\Gamma_*$.

We note that $S\Phi$ lies in $\mathfrak{S}^1(G)$ [cf. § 4.5]. We now start the proof with a very simple observation. If $S\in\mathbb{B}(G)$ satisfies (9-2), then for any $\zeta\in\Gamma\times\Gamma_*$ we obviously have, in abbreviated notation,

$$\int_\Gamma U(\zeta)S\Phi = \int_\Gamma S\Phi = \int_\Gamma \Phi$$

and again

$$\int_\Gamma S^{-1}U(\zeta)S\Phi = \int_\Gamma U(\zeta)S\Phi,$$

so that

$$\psi(\zeta)\int_\Gamma U(\zeta\sigma)\Phi = \int_\Gamma \Phi, \quad \Phi\in\mathfrak{S}^1(G).$$

Hence $\zeta\sigma\in\Gamma\times\Gamma_*$ and $\psi(\zeta) = 1$ $(\zeta\in\Gamma\times\Gamma_*)$, by the lemma. That is to say, S lies in the normalizer of $U(\Gamma\times\Gamma_*)$ in $\mathbb{B}(G)$, the subgroup of all $S\in\mathbb{B}(G)$ such that $S^{-1}U(\Gamma\times\Gamma_*)S = U(\Gamma\times\Gamma_*)$; we denote it by $\mathbb{B}'(G,\Gamma)$. Let $B_o'(G,\Gamma)$ be the subgroup of all $(\sigma,\psi)\in B_o(G)$ such that σ induces an automorphism of $\Gamma\times\Gamma_*$ (without any condition on the Haar modulus!) and ψ is 1 on $\Gamma\times\Gamma_*$; clearly $\pi_o(\mathbb{B}'(G,\Gamma))\subset B_o'(G,\Gamma)$ [π_o as in (1-7)].

The main part of the proof consists in showing that $B_o'(G,\Gamma)$ <u>can be lifted to</u> $\mathbb{B}(G)$, i.e. to $\mathbb{B}'(G,\Gamma)$; this will also show, of course, that $\pi_o(\mathbb{B}'(G,\Gamma)) = B_o'(G,\Gamma)$. The lifting is defined by means of Weil's theta space $L^2(G,\Gamma)$ [§§ 4.7, 4.8]: we put for $(\sigma,\psi)\in B_o'(G,\Gamma)$

$$[\tilde{R}_\Gamma(\sigma,\psi)\theta](z) := \theta(z\sigma)\cdot|\dot{\sigma}|_Q^{1/2}\cdot\psi(z), \quad z\in G\times G^* \ (\theta\in L^2(G,\Gamma)), \tag{9-3}$$

where $\dot{\sigma}$ is the quotient automorphism induced in Q [(7-3)] by σ, in the canonical way (since σ leaves $\Gamma\times\Gamma_*$ invariant, by assumption). We recall that $|\sigma|_{G\times G^*} = 1$ [§ 4.2, Remark 2]; thus [§ 1.6, Remark 2]

$$|\dot{\sigma}|_Q = 1/|\sigma|_{\Gamma\times\Gamma_*}. \tag{9-4}$$

It is readily verified that $\tilde{R}_\Gamma$ in (9-3) is an automorphism of $L^2(G,\Gamma)$ for $(\sigma,\psi)\in B_o'(G,\Gamma)$. It will be noted that $\tilde{R}_\Gamma$ leaves the subspace of continuous functions in $L^2(G,\Gamma)$ invariant, and likewise the subspace $\mathcal{K}(G,\Gamma)$. Moreover, $\tilde{R}_\Gamma$ is a representation of $B_o'(G,\Gamma)$ in $L^2(G,\Gamma)$. We can transfer it to $L^2(G)$ by using the intertwining operator $\tilde{T}_\Gamma$ [(7-4)]: we define

$$R_\Gamma(\sigma,\psi) := \tilde{T}_\Gamma^{-1}\tilde{R}_\Gamma(\sigma,\psi)\tilde{T}_\Gamma. \tag{9-5}$$

Next let us show: $R_\Gamma(\sigma,\psi)$ <u>lies in</u> $\mathbb{B}(G)$, <u>and</u> $\pi_o(R_\Gamma(\sigma,\psi)) = (\sigma,\psi)$. This amounts to

$$U(z)R_\Gamma(\sigma,\psi) = R_\Gamma(\sigma,\psi)\psi(z)U(z\sigma). \tag{9-6}$$

We go back to $L^2(G,\Gamma)$, putting

$$\tilde{U}(z) := \tilde{T}_\Gamma U(z)\tilde{T}_\Gamma^{-1} \in \mathrm{Aut}(L^2(G,\Gamma)).$$

So (9-6) is equivalent to

$$\tilde{U}(z)\tilde{R}_\Gamma(\sigma,\psi) = \tilde{R}_\Gamma(\sigma,\psi)\psi(z)\tilde{U}(z\sigma). \tag{9-7}$$

We can calculate $\tilde{U}(z)$ explicitly:

$$\tilde{U}(z)\theta(z') = \theta(z+z')\cdot F(z',z), \quad z' \in G\times G^*.$$

Then we can verify (9-7) by using the addition formula (1-3). Now for every $S \in R_\Gamma(B_0'(G,\Gamma))$ we have by (9-5) and the definition of $\tilde{R}_\Gamma(\sigma,\psi)$ [(9-3)], if we take there $z = (0,0) \in G\times G^*$,

$$\int_\Gamma S\Phi = |\dot{\sigma}|_Q^{1/2}\cdot\int_\Gamma \Phi, \quad \Phi \in \mathcal{S}^1(G).$$

Thus the Poisson-Weil formula holds for $S \in R_\Gamma(B_0(G,\Gamma))$, since by definition $|\sigma|_{\Gamma\times\Gamma_*} = 1$ for $(\sigma,\psi) \in B_0(G,\Gamma)$, whence $|\dot{\sigma}|_Q = 1$ by (9-4). On the other hand, if $S \in \mathbb{B}(G)$ satisfies (9-2), then (as we saw at the beginning of the proof) S must be in $\mathbb{B}'(G,\Gamma)$, so $\pi_0(S) = (\sigma,\psi)$ lies in $B_0'(G,\Gamma)$. This gives $S = c\cdot R_\Gamma(\sigma,\psi)$ with $c \in T$ [§ 4.6], and hence

$$\int_\Gamma \Phi = \int_\Gamma S\Phi = c\cdot|\dot{\sigma}|_Q^{1/2}\cdot\int_\Gamma \Phi, \quad \Phi \in \mathcal{S}^1(G),$$

so that $c = 1$, $|\dot{\sigma}|_Q = 1$, and in turn $|\sigma|_{\Gamma\times\Gamma_*} = 1$. The lifting R_Γ being defined by (9-5), we have, finally,

$$\mathbb{B}(G,\Gamma) = R_\Gamma(B_0(G,\Gamma)), \tag{9-8}$$

and the proof is complete.

REMARKS.

1. We call the lifting - in fact, representation - R_Γ described in the theorem above the <u>theta lifting</u>, or <u>theta representation</u>, <u>of</u> $B_0(G,\Gamma)$. We note the relation

$$\pi_0(\mathbb{B}(G,\Gamma)) = B_0(G,\Gamma).$$

2. If $\Gamma\times\Gamma_*$ is discrete or compact, i.e. if Γ is discrete and G/Γ compact, or if Γ is compact and open, then $|\sigma|_{\Gamma\times\Gamma_*}$ is always 1. Here the distinction between $\mathbb{B}(G,\Gamma)$ and $\mathbb{B}'(G,\Gamma)$, and between $B_0(G,\Gamma)$ and $B_0'(G,\Gamma)$, disappears. This distinction is omitted altogether in Weil's memoir; cf. the remarks in his Coll. Papers [27, vol.III, p.445].

3. In the important special case of a <u>compact and open subgroup</u> Γ, let Φ_Γ be the characteristic function of Γ on G, i.e. 1 on Γ, 0 elsewhere on G. We note that Φ_Γ is in $\mathcal{S}^1(G)$. It is readily seen that

$\tilde{T}_\Gamma \Phi_\Gamma = \Phi_{\Gamma\times\Gamma_*}$, which is obviously invariant under $\tilde{R}_\Gamma(\sigma,\psi)$ for (σ,ψ) in $B_0(G,\Gamma)$. It follows that Φ_Γ <u>is invariant under</u> $R_\Gamma(\sigma,\psi)$, $(\sigma,\psi) \in B_0(G,\Gamma)$. This property will be essential in § 9, in connexion with the metaplectic group in the adelic case.

4. When $\Gamma = \{0\}$, $\mathbb{B}(G,\Gamma)$ is the group of all operators $M(\alpha)A(\psi_G)$, $\alpha \in Aut(G)$, $\psi_G \in Ch_2(G)$; for $\Gamma = G$ it is the group obtained through the intertwining Fourier operator from the analogous operators in $L^2(G^*)$.

5. <u>Let</u> Γ <u>be a discrete subgroup of</u> G <u>with</u> G/Γ <u>compact</u>, so that Γ_* is also discrete. Suppose that there exists an isomorphism γ of G^* onto G which maps Γ_* onto Γ. We normalize the Haar measures on G and G^* by stipulating that G/Γ and G^*/Γ_* have Haar measure 1, and taking the ordinary sum on Γ and Γ_*; so the Haar measures on G, G^* are dual. The Haar modulus $|\gamma|$ is 1, as is readily seen. If such a γ exists, then, as the theorem above, or a direct calculation, shows, W^γ <u>is in</u> $\mathbb{B}(G,\Gamma)$ (and, of course, conversely); this is the really interesting case. <u>In this case the Poisson-Weil formula reduces to the classical Poisson summation formula</u>. <u>Analogously if</u> Γ <u>is a compact, open subgroup</u>. The classical examples are $G = X$, a finite-dimensional vector space over $\mathbb{R}$, Γ a lattice in X; note that in this case we can always obtain an ($\mathbb{R}$-linear) isomorphism γ of X^* onto X mapping Γ_* onto Γ. Likewise if X is a finite-dimensional vector space over the p-adic numbers $\mathbb{Q}_p$, X° a basis of X, Γ the compact and open subgroup of X generated by X° over the p-adic integers $\mathbb{Z}_p$. Other cases will be considered later [§§ 7.4, 9.1, 9.2]. An explicit determination of (generators of) $\mathbb{B}(\mathbb{R}^n,\mathbb{Z}^n)$ is contained in Igusa's book [10, Ch. I, § 10]; analogously one can consider $\mathbb{B}(\mathbb{Q}_p{}^n,\mathbb{Z}_p{}^n)$.

6. The opening sentence in Weil's memoir [25], about Siegel's theta series and the symplectic group, may well be applied to the space $L^2(G,\Gamma)$ of Weil's theta functions and to the 'theta representation' in $L^2(G,\Gamma)$ which is the basis of Weil's theorem 4. It may be added that the theta representation is not an induced representation in the sense of Mackey; for connexions with Mackey's theory see Mackey's work [12, 13, 14] and Igusa's book [10, Ch. I, § 5].

<u>4.10</u> We now introduce a topology in $B_0(G)$. We provide $Sp(G)$ with the automorphism topology, as a subgroup of $Aut(G\times G^*)$, and $Ch_2(G\times G^*)$ with the topology of uniform convergence on compact sets. Then we can form the semi-direct product $Sp(G)\times_s Ch_2(G\times G^*)$, with the law of multiplication (1-6) and the product topology. $B_0(G)$ is a subgroup and we may speak of the <u>product topology in</u> $B_0(G)$.

Now $B_o(G)$ is also a topological group as a subgroup of $Aut(A(G))$. We show: <u>the automorphism topology of</u> $B_o(G)$ <u>coincides with the product topology</u>.

First, choose a nd. $\mathcal{U}$ of the neutral element $(1,1)$ in $B_o(G)$ for the automorphism topology. $\mathcal{U}$ is defined as follows: (i) let a compact set in $A(G)$ be given which we may take, without loss of generality, of the form $K\times T$, with K a compact set in $G\times G^*$; (ii) let a nd. of the neutral element $(0,1)$ of $A(G)$ be given which we may assume to be $V\times U_\varepsilon(1)$, with V a nd. of 0 in $G\times G^*$ and

$$U_\varepsilon(1) := \{ t \mid t\in T,\ |t-1| < \varepsilon \}.$$

Then the nd. $\mathcal{U}$ is defined as

$$\mathcal{U} := \mathcal{U}_{B_o(G)}(K\times T, V\times U_\varepsilon(1)) \tag{10-1}$$

and consists of all $(\sigma,\psi)\in B_o(G)$ such that, for $\alpha := (\sigma,\psi)$ and $\alpha := (\sigma,\psi)^{-1}$

$$(z,t)^{\alpha}\cdot(z,t)^{-1} \in V\times U_\varepsilon(1) \text{ for all } (z,t)\in K\times T.$$

We want to show: if $(\sigma,\psi)\in B_o(G)$ is such that σ is near to 1 in $Sp(G)$ and ψ is near to 1 in $Ch_2(G\times G^*)$, then (σ,ψ) is in $\mathcal{U}$. For this, we note that

$$(z,t)^{(\sigma,\psi)}\cdot(z,t)^{-1} = (z\sigma-z, F(z-z\sigma,z)\cdot\psi(z)) \tag{10-2a}$$

$$(z,t)^{(\sigma,\psi)^{-1}}\cdot(z,t)^{-1} = (z\sigma^{-1}-z, F(z-z\sigma^{-1},z)\cdot\psi(z\sigma^{-1})^{-1}). \tag{10-2b}$$

Since F is a bicharacter of $(G\times G^*)\times(G\times G^*)$, there is for K as above and $\varepsilon > 0$ a compact nd. V' of 0 in $G\times G^*$ such that [§ 1.3]

$$|F(z',z) - 1| < \varepsilon/2 \quad \text{for all } z'\in V' \text{ and all } z\in K. \tag{10-3}$$

We show: if $(\sigma,\psi)\in B_o(G)$ satisfies (with an obvious notation)

$$\sigma\in \mathcal{U}_{Aut(G\times G^*)}(K, V\cap V'),\ \psi\in U_{Ch_2(G)}(K+V', \varepsilon/2), \tag{10-4}$$

then $(\sigma,\psi)\in\mathcal{U}$.

In view of (10-2), we have to show: if (10-4) holds, then for all z in K

$$z\sigma-z \text{ and } z\sigma^{-1}-z \text{ lie in } V, \tag{10-5a}$$

$$|F(z-z\sigma,z)\cdot\psi(z) - 1| < \varepsilon,\ |F(z-z\sigma^{-1},z)\cdot\psi(z\sigma^{-1})^{-1} - 1| < \varepsilon. \tag{10-5b}$$

Now the first condition in (10-4) says that $z\sigma-z$ and $z\sigma^{-1}-z$ lie in $V\cap V'$, for all $z\in K$, so (10-5a) certainly holds; thus, by the choice of V' we have, since in (10-3) we may replace z' by $-z'$,

$$|F(z-z\sigma,z) - 1| < \varepsilon/2,\ |F(z-z\sigma^{-1},z) - 1| < \varepsilon/2 \text{ for } z\in K.$$

Moreover, by the second part of (10-4),

$$|\psi(z) - 1| < \varepsilon/2, \ |\psi(z\sigma^{-1}) - 1| < \varepsilon/2 \text{ for } z \in K$$

[note that $z\sigma^{-1} \in K+V'$]. It follows that (10-5b) holds. Thus, given $\mathcal{U}$ as in (10-1), we shall have $(\sigma,\psi) \in \mathcal{U}$ if (10-4) holds.

Secondly, choose a nd. $\mathcal{U}_{\mathrm{Aut}(G\times G^*)}(K,V)$ of e in Aut(G×G*) and a nd. $U(K,\varepsilon)$ of 1 in $Ch_2(G\times G^*)$ (there is of course no loss of generality in choosing the same compact set $K \subset G\times G^*$ for both nds.). Take a compact nd. V' of 0 in G×G* such that (10-3) holds. Then, if

$$(\sigma,\psi) \in \mathcal{U}_{B_o(G)}(K\times T, (V\cap V')\times U_{\varepsilon/2}(1)),$$

we shall have [cf. (10-2a)]

$$\sigma \in \mathcal{U}_{\mathrm{Aut}(G\times G^*)}(K, V\cap V') \subset \mathcal{U}_{\mathrm{Aut}(G\times G^*)}(K,V) \text{ and } \psi \in U(K,\varepsilon),$$

the second relation stemming from

$$|\psi(z) - 1| \le |\psi(z)\cdot F(z-z\sigma,z) - 1| + |1 - F(z-z\sigma,z)|$$
$$\le \varepsilon/2 + \varepsilon/2 \text{ for all } z \in K \text{ [cf. (10-2a), (10-3)]}.$$

This concludes the proof.

4.11 <u>Suppose that</u> $x \longmapsto 2x$ <u>is an automorphism of</u> G. Then $z \longmapsto 2z$ is an automorphism of G×G* [cf. § 1.4 (vii)]. Thus we can define, for $\sigma \in Sp(G)$, and with F as in § 1 (8-2),

$$\psi_\sigma(z) := F(2^{-1}z\sigma, z\sigma) = F(z\sigma, 2^{-1}z\sigma), \quad z \in G\times G^*. \tag{11-1}$$

Then ψ_σ is in $Ch_2(G\times G^*)$ and belongs to σ, i.e. (1-3) holds; so (σ,ψ_σ) is in the group $B_o(G)$ of § 4.1. The elements (σ,ψ_σ), $\sigma \in Sp(G)$, form a subgroup, say $B_*(G)$, as is readily seen. Here, of course, $\sigma \longmapsto \psi_\sigma$ is a <u>continuous</u> mapping of Sp(G) into $Ch_2(G\times G^*)$, thus $B_*(G)$ is isomorphic to Sp(G) as a <u>topological</u> group [§ 1.12].

By a short calculation we obtain from (11-1), after some simplifications, for $\sigma = \begin{pmatrix}\alpha & \beta\\ \gamma & \delta\end{pmatrix} \in Sp(G)$

$$\psi_\sigma(x,x^*) = \langle 2^{-1}x\alpha, x\beta\rangle\cdot\langle x^*\gamma, x\beta\rangle\cdot\langle 2^{-1}x^*\gamma, x^*\delta\rangle, \ (x,x^*) \in G\times G^*. \tag{11-2}$$
$$= \langle x, 2^{-1}x\beta\alpha^*\rangle\cdot\langle x^*\gamma, x\beta\rangle\cdot\langle 2^{-1}x^*\gamma\delta^*, x^*\rangle$$

(Weil [25, n° 5, pp. 150-151]). We know that $\beta\alpha^*$, $\gamma\delta^*$ are symmetric [(2-3)]. From (11-2) we obtain, if we put successively

$$\sigma := \begin{pmatrix}\alpha & 0\\ 0 & \alpha^{*-1}\end{pmatrix}, \ \begin{pmatrix}1 & \beta\\ 0 & 1\end{pmatrix}, \ \begin{pmatrix}0 & -\gamma^{*-1}\\ \gamma & 0\end{pmatrix},$$

with $\alpha \in \mathrm{Aut}(G)$, $\beta \in \mathrm{Mor}(G,G^*)$, $\beta = \beta^*$, $\gamma \in \mathrm{Is}(G^*,G)$, and also recall the notations (1-8a,b,c), that

$$(\sigma,\psi_\sigma) = m_0(\alpha),\ a_0(\psi_\beta),\ w_0^\gamma \quad [\psi_\beta(x) := \langle x, 2^{-1}x\beta\rangle].$$

Now consider the normal subgroup $A_0(G)$ of $B_0(G)$ introduced in § 4.2, Remark: with the topology given to $B_0(G)$, $A_0(G)$ is isomorphic, as a topological group, to $G\times G^*$. Moreover, $B_0(G)$ is the <u>topological semi-direct product of</u> $B_*(G)$ <u>and</u> $A_0(G)$:

$$B_0(G) = A_0(G)\cdot B_*(G), \quad A_0(G)\cap B_*(G) = \{(1,1)\}, \tag{11-4}$$

and the topology of $B_0(G)$ coincides with that of $A_0(G)\times B_*(G)$. For, in the splitting

$$(\sigma,\psi) = (1, \chi_{a*}\otimes\bar{\chi}_a)\cdot(\sigma,\psi_\sigma), \tag{11-5}$$

- where it is convenient, for later applications, to put a bar over χ_a to obtain agreement with formula (1-10) - the second factor on the right in (11-5) depends continuously [on σ, and thus] on (σ,ψ), hence so does the first, and the assertion follows.

REMARK. If $x \longmapsto 2x$ is an automorphism of G, it is thus clear that the morphism $(\sigma,\psi) \longmapsto \sigma$ of $B_0(G)$ onto Sp(G) is strict (open), i.e. a homomorphism. This is still true for general l.c.a. G, but far more difficult to prove (Igusa [9, Prop.2, p.203]); we shall not need it.

§5. *Vector spaces and quadratic forms over local fields*

5.1 A commutative locally compact field which is locally compact, but not discrete, is called a local field. We denote such a field by $\mathbb{K}$. The basic structure theorem states that the possible types of $\mathbb{K}$ are as follows.

(I) If $\mathbb{K}$ is connected, then $\mathbb{K}$ is $\mathbb{R}$ or $\mathbb{C}$.

(II) Otherwise we have: (a) If $\mathbb{K}$ is of characteristic 0, then $\mathbb{K}$ is the field of p-adic numbers $\mathbb{Q}_p$ (p a prime) or a finite algebraic extension of $\mathbb{Q}_p$. (b) If $\mathbb{K}$ is of characteristic $p > 0$, then $\mathbb{K}$ is (isomorphic to) the field $\mathbb{F}_q((T))$ of all 'formal' power series in an 'indeterminate' T with coefficients in a finite field $\mathbb{F}_q$, where q, the number of elements, is a power of p. A formal power series contains by definition at most finitely many negative powers of T (there should be no confusion with the other use of T [§ 1.1]).

In case (I) the topology is defined by the classical absolute value. In case (IIa) the topology of $\mathbb{Q}_p$ is defined by the familiar p-adic absolute value $|\ |_p$; the topology (and the absolute value) carries over to any finite algebraic extension (which is a vector space of finite dimension over $\mathbb{Q}_p$). In case (IIb) the topology is defined by the following absolute value: $|x| := q^{-N}$ if $x = T^N \cdot \sum_{n\geq 0} c_n T^n$, $c_0 \neq 0$ (N in $\mathbb{Z}$), $|0| := 0$.

In case (II) the absolute value is ultrametric, i.e.

$$|x+y| \leq \max\{|x|, |y|\}.$$

Then $\mathbb{K}$ itself is also called ultrametric.

We may define $|x|$ by the Haar modulus of the automorphism $t \longmapsto xt$, $t \in \mathbb{K}$ (for any Haar measure on the additive group of $\mathbb{K}$), $x \neq 0$: let

$$\mathrm{mod}_{\mathbb{K}}(x) \cdot \int_{\mathbb{K}} f(xt)dt = \int_{\mathbb{K}} f(t)dt, \quad f \in \mathcal{K}(\mathbb{K}), \quad \text{if } x \neq 0,$$

and put $\mathrm{mod}_{\mathbb{K}}(0) := 0$; then

$$|x| = \mathrm{mod}_{\mathbb{K}}(x), \ x \in \mathbb{K} \neq \mathbb{C}, \quad |x| = \sqrt{\{\mathrm{mod}_{\mathbb{K}}(x)\}} \text{ if } \mathbb{K} = \mathbb{C}.$$

If $\mathbb{K}$ is ultrametric, we put

$$\mathfrak{o} = \mathfrak{o}_{\mathbb{K}} := \{x \mid x \in \mathbb{K}, |x| \leq 1\},$$

$$\mathfrak{p} = \mathfrak{p}_{\mathbb{K}} := \{x \mid x \in \mathbb{K}, |x| < 1\}. \qquad (1\text{-}1)$$

Here $\mathfrak{o}$ is a compact ring and $\mathfrak{p}$ is the unique maximal ideal of $\mathfrak{o}$. Moreover, $\mathfrak{o}/\mathfrak{p}$ is a finite field $\mathbb{F}_q$, where q is a power of the prime p. There is a $\pi \in \mathfrak{o}$ such that $\mathfrak{p} = \pi \cdot \mathfrak{o}$, i.e. $\mathfrak{p} = (\pi)$; for any such π we put

$$\mathfrak{p}^N := (\pi^N) = \pi^N \cdot \mathfrak{o} = \{x \mid x \in \mathbb{K},\ |x| \le q^{-N}\},\ N \ge 1. \tag{1-2}$$

For $\mathbb{K} = \mathbb{F}_q((T))$ [case (IIb) above] we have $\mathfrak{o} = \mathbb{F}_q[[T]]$, the ring of 'integral' power series in T (i.e., containing no negative powers of T) and we can take $\pi = T$.

For $x \in \mathbb{K}$, $x \ne 0$, we put

$$\mathrm{ord}(x) = \mathrm{ord}_{\mathbb{K}}(x) := m \text{ if } |x| = q^{-m}. \tag{1-3}$$

REFERENCES. Weil [25, n° 24, p.172], [26, Ch.I], Cassels-Fröhlich [6, Ch.II], Serre [23, Ch.II], Bourbaki [1, Chap.6, § 9, n° 3]. There seems to be no generally adopted terminology and notation.

5.2 Let X be a vector space of finite dimension, say n, over the local field $\mathbb{K}$. It will be convenient for later purposes to use a basis in X, say $X^\circ = \{e_1, \dots, e_n\}$, so we can identify X with $\mathbb{K}^n$. It will also be convenient to use the following norm in $\mathbb{K}^n$: *for* $x = (x_1, \dots, x_n)$ *in* $\mathbb{K}^n$ *we put*

$$\|x\| := \max\{|x_j|,\ 1 \le j \le n\} \text{ if } \mathbb{K} \text{ is ultrametric}, \tag{2-1a}$$

$$\|x\| := n \cdot \max\{|x_j|,\ 1 \le j \le n\} \text{ if } \mathbb{K} \text{ is } \mathbb{R} \text{ or } \mathbb{C}. \tag{2-1b}$$

Denote by $M_n(\mathbb{K})$ the algebra (over $\mathbb{K}$) of all n×n matrices $\alpha = (a_{ij})$, $a_{ij} \in \mathbb{K}$, $1 \le i, j \le n$ (the notation α, already used for automorphisms of groups [§ 1.6], should not cause difficulty, since the meaning will be clear from the context). We introduce the following norms in $M_n(\mathbb{K})$: *for* $\alpha = (a_{ij})_{1 \le i, j \le n}$ *we put*

$$\|\alpha\| := \max\{|a_{ij}|,\ 1 \le i, j \le n\} \text{ if } \mathbb{K} \text{ is ultrametric}, \tag{2-2a}$$

$$\|\alpha\| := n \cdot \max\{|a_{ij}|,\ 1 \le i, j \le n\} \text{ if } \mathbb{K} \text{ is } \mathbb{R} \text{ or } \mathbb{C}. \tag{2-2b}$$

The usefulness of the particular norms (2-1), (2-2) for our purposes will appear later [§ 6]. With the norms above, $M_n(\mathbb{K})$ is a Banach algebra over $\mathbb{K}$ (the unit matrix has norm n in case (2-2b), but this will not matter). We have

$$\|x\alpha\| \le \|\alpha\| \cdot \|x\|,\ x \in \mathbb{K}^n,\ \alpha \in M_n(\mathbb{K}), \tag{2-3}$$

- we multiply by α on the right, x being a row vector - and

$$|x \cdot {}^t y| \le \|x\| \cdot \|y\|,\quad x, y \text{ in } \mathbb{K}^n, \tag{2-4}$$

where ${}^t y$ denotes the transposed vector, so $x \cdot {}^t y$ is the scalar product of x and y.

Between the norm (2-2) and the operator norm in $M_n(\mathbb{K})$, defined by

$$\|\alpha\|_o := \max\{\|x\alpha\| \mid x\in \mathbb{K}^n,\ \|x\| \le 1\},\quad \alpha\in M_n(\mathbb{K}), \tag{2-5}$$

the following relation holds:

$$\|\alpha\|_o = \|\alpha\|,\ \alpha\in M_n(\mathbb{K}),\ \text{if } \mathbb{K} \text{ is ultrametric}, \tag{2-6a}$$

$$\|\alpha\|_o \le \|\alpha\| \le n\cdot\|\alpha\|_o,\ \alpha\in M_n(\mathbb{K}),\ \text{if } \mathbb{K} \text{ is } \mathbb{R} \text{ or } \mathbb{C}. \tag{2-6b}$$

To show this, note first that by (2-3) we have in both cases

$$\|\alpha\|_o \le \|\alpha\|. \tag{2-7}$$

Next, let $\mathbb{K}$ be ultrametric. Then we consider the vectors

$$x^{(j)} := (\delta_{ij})_{1\le i\le n} \text{ for } j = 1,\dots,n \text{ (Kronecker delta)}. \tag{2-8}$$

We have $\|x^{(j)}\| = 1$ and $x^{(j)}\alpha = a^{(j)}$, the vector in row j of α; thus

$$\|a^{(j)}\| \le \|\alpha\|_o, \tag{2-9}$$

i.e. $\sup\{|a_{jk}|,\ 1 \le k \le n\} \le \|\alpha\|_o$ for each j, $j = 1,\dots,n$. Now this yields $\|\alpha\| \le \|\alpha\|_o$, and in view of (2-7) the equality (2-6a) follows. Finally, let $\mathbb{K}$ be $\mathbb{R}$ or $\mathbb{C}$. Then the vector $x^{(j)}$ in (2-8) has norm n, and we obtain in place of (2-9) the inequality $\|a^{(j)}\| \le n\cdot\|\alpha\|_o$. Hence $\|\alpha\| \le n\cdot\|\alpha\|_o$ and this, combined with (2-7), yields (2-6b).

Let $\alpha\in M_n(\mathbb{K})$ be such that

$$\|\alpha\| \le \varepsilon \text{ with } 0 < \varepsilon < 1.$$

Then $(1_n - \alpha)^{-1}$, where 1_n denotes the unit matrix, exists and

$$(1_n - \alpha)^{-1} = 1_n + \alpha',\ \text{with } \alpha' := \sum_{n\ge 1}\alpha^n. \tag{2-10}$$

Replacing here α by $1_n - \alpha$, we obtain: <u>if</u> $\|\alpha - 1_n\| \le \varepsilon < 1$, <u>then</u> α^{-1} <u>exists; moreover</u>,

$$\|\alpha^{-1} - 1_n\| \le \varepsilon \text{ if } \mathbb{K} \text{ is ultrametric}, \tag{2-11a}$$

$$\|\alpha^{-1} - 1_n\| \le \varepsilon/(1-\varepsilon) \text{ if } \mathbb{K} \text{ is } \mathbb{R} \text{ or } \mathbb{C}. \tag{2-11b}$$

Thus not only multiplication, but also the inverse is continuous on $GL_n(\mathbb{K})$, the group of all invertible matrices in $M_n(\mathbb{K})$, i.e. $GL_n(\mathbb{K})$ <u>is a topological group</u> in the topology induced by that of $M_n(\mathbb{K})$, the norm topology, and is locally compact.

REMARK. The relation (2-10) shows for <u>ultrametric</u> $\mathbb{K}$ that, if U is a <u>compact subring</u> of $M_n(\mathbb{K})$ such that $\|\alpha\| < 1$ for all $\alpha\in U$, then $1_n + U$ is a compact <u>subgroup</u> of $GL_n(\mathbb{K})$, open if U is a nd. of 0 in $M_n(\mathbb{K})$.

<u>5.3</u> We discuss here some simple properties of local fields; later an extension to higher dimension will be given [§ 6]. In the ultra-

metric case we use the notation $\mathfrak{p}$ and $\mathfrak{p}^N$ [(1-1),(1-2)]; for $\mathbb{K} = \mathbb{R}$ or $\mathbb{C}$ we introduce the balls

$$B(\varepsilon;\mathbb{K}) := \{t \mid t \in \mathbb{K},\ |t| < \varepsilon\}, \qquad (3\text{-}1a)$$

$$B(1,\varepsilon;\mathbb{K}) := \{t \mid t \in \mathbb{K},\ |t-1| < \varepsilon\} = 1 + B(\varepsilon;\mathbb{K}). \qquad (3\text{-}1b)$$

LEMMA. _Let_ $\mathbb{K}$ _be a local field._

(i) _If_ $\mathbb{K}$ _is ultrametric and_ $c \in \mathbb{K}$, $|c| = 1$, _then the mapping_ $t \longmapsto s$, $s := t^2 + c\cdot t$, _maps_ $\mathfrak{p}^N$ _bijectively onto itself_, $N \geq 1$.

(ii) _If_ $\mathbb{K}$ _is_ $\mathbb{R}$ _or_ $\mathbb{C}$, _the mapping_ $t \longmapsto s$, $s := t^2 + 2t$, _maps_ $B(1;\mathbb{K})$ _injectively into_ $\mathbb{K}$ _and, for_ $0 < \varepsilon \leq 1$, _the image of_ $B(\varepsilon;\mathbb{K})$ _contains_ $B(\varepsilon;\mathbb{K})$ [cf. (3-1a)].

Proof of (i). Consider first $\mathfrak{p}$, i.e. let $N = 1$. If $t \in \mathfrak{p}$, i.e. $|t| < 1$, then we have

$$|s| = |t|\cdot|t+c| = |t| \qquad (3\text{-}2)$$

[here $|t+c| = |c|$ if $|t| < |c|$, since $|c| \leq \max\{|c+t|, |-t|\}$]. Thus $\mathfrak{p}$ is mapped into $\mathfrak{p}$. Now let us show that, conversely, given $s \in \mathfrak{p}$, the equation $s = t^2 + c\cdot t$ for t has precisely one solution in $\mathfrak{p}$. We put

$$t_0 := s,\quad t_n := s/(c+t_{n-1}),\quad n > 0. \qquad (3\text{-}3)$$

Then $|t_n| = |s| < 1$, by induction. Next we have

$$t_{n+1} - t_n = s\cdot(t_{n-1} - t_n)/[(c+t_n)\cdot(c+t_{n-1})],\quad n \geq 1,$$

and here $|c+t_n| = 1$ for $n \geq 0$, since $|c| = 1$, $|t_n| < 1$; so

$$|t_{n+1} - t_n| = |s|\cdot|t_n - t_{n-1}|,\quad n \geq 1,$$

whence

$$|t_{n+1} - t_n| = |s|^n\cdot|t_1 - t_0|,\quad n \geq 0.$$

Thus $t_0 + \sum_{n\geq 0}(t_{n+1} - t_n)$ converges in $\mathbb{K}$, say to t_*. Hence $\lim_{n\to\infty} t_n = t_*$ exists and [cf. (3-3)] satisfies $s = t_*^2 + c\cdot t_*$; moreover, we have $|t_*| = |s| < 1$, i.e. t is in $\mathfrak{p}$. Also, the equation $s = t^2 + c\cdot t$ has obviously only one solution $t_* \in \mathfrak{p}$, the other one being $-(c+t_*)$ which is not in $\mathfrak{p}$. Thus the case $N = 1$ of part (i) is proved. Then the relation (3-2) above shows that (i) holds, in fact, for $\mathfrak{p}^N$, $N \geq 1$.

Proof of (ii). Here the method is quite analogous. Consider $B(\varepsilon;\mathbb{K})$ for $\varepsilon = 1$: there can be at most one $t \in B(1;\mathbb{K})$ satisfying $s = t^2 + 2t$, since the sum of the two solutions is -2. To show the existence of the solution, define for given $s \in B(1;\mathbb{K})$ the sequence $(t_n)_{n\geq 0}$ as in (3-3), with c replaced by 2. This gives $|t_n| \leq |s|$ by induction, and

$$|t_{n+1}-t_n| \le |s|^n \cdot |t_1-t_0|, \ n \ge 0.$$

It follows as before that $\lim t_n =: t_* \ (n \longrightarrow \infty)$ exists and satisfies $s = t_*^2+2t_*$, $|t_*| \le |s|$. The assertion in part (ii) concerning $B(\varepsilon;\mathbb{K})$ results, first for $\varepsilon = 1$, then for general ε, $0 < \varepsilon \le 1$.

The above proof shows the analogy between the ultrametric and the classical case. For ultrametric $\mathbb{K}$ the result can also be shown by means of Hensel's lemma (cf. Serre [23, Prop.7, p.23]), as Professor T. A. Springer has kindly pointed out to the author.

<u>5.4</u> From the lemma in § 5.3 we obtain the following corollary which we also state as a lemma, for later reference.

LEMMA. <u>Let</u> $\mathbb{K}$ <u>be a local field of characteristic</u> $\neq 2$.

(i) <u>If</u> $\mathbb{K}$ <u>is ultrametric, the mapping</u> $x \longmapsto x^2$ <u>gives a bijection, hence a homeomorphism, of</u> $1+\mathfrak{p}^N$ <u>onto</u> $1+\mathfrak{p}^{N+\mathrm{ord}(2)}$ <u>for all</u> $N > \mathrm{ord}(2)$ [cf. (1-3) and note that $\mathrm{ord}(2) > 0$ if, and only if, $\mathbb{K}$ is $\mathbb{Q}_2$ or an algebraic extension of $\mathbb{Q}_2$, i.e. in the case (IIa) of § 5.1, with $p = 2$].

(ii) <u>If</u> $\mathbb{K}$ <u>is</u> $\mathbb{R}$ <u>or</u> $\mathbb{C}$, <u>then the mapping</u> $x \longmapsto x^2$ <u>gives an injection of</u> $B(1,1;\mathbb{K}) = 1 + B(1;\mathbb{K})$ [(3-1b)] <u>into</u> $\mathbb{K}$, <u>and for</u> $0 < \varepsilon \le 1$ <u>the image of</u> $B(1,\varepsilon;\mathbb{K})$ <u>contains</u> $B(1,\varepsilon;\mathbb{K})$.

To prove (i), put $x = 1+u$ and consider the mapping $u \longmapsto u^2+2u$. This maps $2\cdot\mathfrak{p}^N$ bijectively onto $4\cdot\mathfrak{p}^N$: for, let us put $u = 2t$, for t in $\mathfrak{p}^N$; by part (i) of the lemma in § 5.3, $2t \longmapsto 4(t^2+t)$ is a bijection of $2\cdot\mathfrak{p}^N$ onto $4\cdot\mathfrak{p}^N$. Now $2\cdot\mathfrak{p}^N = \mathfrak{p}^{N+\mathrm{ord}(2)}$, $4\cdot\mathfrak{p}^N = \mathfrak{p}^{N+2\cdot\mathrm{ord}(2)}$; then we write N instead of $N+\mathrm{ord}(2)$, whence $N > \mathrm{ord}(2)$, and obtain (i).

Statement (ii) is simply a slightly different formulation of part (ii) of the lemma in § 5.3.

<u>5.5</u> Let $\mathbb{K}$ be a local field of characteristic 2. Then $\mathbb{K}$ is (isomorphic to) a field $\mathbb{F}_q((T))$, $q = 2^r$ [§ 5.1]. The equation $x^2 = c$ has at most one solution in $\mathbb{K}$. Put, for a moment,

$$\mathbb{K}^2 := \{c^2 \mid c \in \mathbb{K}\}.$$

$\mathbb{K}^2$ is a subfield of $\mathbb{K}$ and consists of all c of the form

$$c = T^{2m} \cdot \sum_{k\ge0} c_k T^{2k}, \quad m \in \mathbb{Z}, \ c_k \in \mathbb{F}_q.$$

Indeed, since $q-1$ is odd, the equation $x^2 = c_k$ in $\mathbb{F}_q$ has exactly one solution for each $k \ge 0$, say a_k, so $c = (T^m \cdot \sum_{k\ge0} a_k T^k)^2$. The field $\mathbb{K}$ may be considered as a quadratic extension of $\mathbb{K}^2$. This will be useful later, so we state it as a lemma, for reference.

LEMMA. <u>Let</u> $\mathbb{K}$ <u>be a local field of characteristic</u> 2. <u>For each</u> $a \in \mathbb{K}$ <u>there are unique elements</u> u, v <u>in</u> $\mathbb{K}$ <u>such that</u>

$$a = u^2 + T\cdot v^2,$$

<u>and if</u> a <u>is in</u> $\mathfrak{p}^{2N}$, <u>then</u> u <u>and</u> v <u>are in</u> $\mathfrak{p}^N$, <u>and conversely</u>.

The uniqueness is clear, since T is not in $\mathbb{K}^2$. Now write

$$a = \sum_{k\geq k_o} a_k T^k = a' + T\cdot a'',$$

with

$$a' := \sum_{k'}{}^{k' \text{ even}}_{k' \geq k_o} a_{k'} T^{k'}, \quad a'' := \sum_{k''}{}^{k'' \text{ odd}}_{k'' \geq k_o} a_{k''} T^{k''-1}.$$

Thus a', a" are in $\mathbb{K}^2$. If $a \in \mathfrak{p}^{2N}$, we may put $k_o := 2N$; then $a' \in \mathfrak{p}^{2N}$, $a'' \in \mathfrak{p}^{2N}$ and $a' = u^2$, $a'' = v^2$, with u, v in $\mathfrak{p}^N$, and conversely.

REMARK. This lemma shows, in particular: when char $\mathbb{K} = 2$, then the additive group of $\mathbb{K}$ is an example of a l.c.a. group G such that G is isomorphic, as a topological group, to G×G.

It will be our next task to provide an appropriate extension, for later applications, of the lemmas in §§ 5.4, 5.5 to a vector space of finite dimension over $\mathbb{K}$. For this, we need some preparation.

<u>5.6</u> A quadratic form on a finite-dimensional vector space X over a local field $\mathbb{K}$ may be defined, after Bourbaki, as a function f: $X \longrightarrow \mathbb{K}$ such that

(i) $f(cx) = c^2\cdot f(x)$, $x \in X$, $c \in \mathbb{K}$,

(ii) $f(x+y) = f(x) + f(y) + b_f(x,y)$, where b_f is a $\mathbb{K}$-bilinear form on X×X.

[Cf. Bourbaki, Algèbre, Chap. 9 (1959), § 3, n° 4, p. 54, and foot-note (****), p. 189.]

If an isomorphism of X onto $\mathbb{K}^n$ is given by means of a basis $X^\circ = (e_j)_{1\leq j\leq n}$, (n := dim X), then, as is readily seen,

$$f(x) = \sum_{1\leq i\leq j\leq n} f_{ij}x_i x_j, \; f_{ij} \in \mathbb{K}, \text{ for } x = \sum_{j=1}^{n} x_j e_j.$$

If char $\mathbb{K} \neq 2$, we can write this in matrix notation:

$$f(x) = x\alpha\, {}^t x, \quad \alpha = (a_{ij}), \tag{6-1}$$

where $a_{ij} = a_{ji} := 2^{-1} f_{ij}$, $i \neq j$, $a_{ii} := f_{ii}$, $1 \leq i, j \leq n$.

The quadratic forms on X constitute a vector space Q(X), of dimension n(n+1)/2 over $\mathbb{K}$. We write $Q_n(\mathbb{K})$ for $Q(\mathbb{K}^n)$, $\mathbb{K}^n$ having the canonical basis. $Q(X)$ is even a Banach space over $\mathbb{K}$: we can take as norm

$$\|f\|_{Q(X)} := \max \{|f(x)| \mid \|x\| \le 1\}.$$

It should be noted that the topology of $Q(X)$ is that of uniform convergence on compact sets of X. We can also use the norm (depending on the basis X°)

$$\|f\|'_{Q(X)} = \max_{1\le i\le j\le n} |f_{ij}|.$$

These norms are equivalent, in a familiar sense.

If char $\mathbb{K} \ne 2$, we can identify $Q(X)$ by means of (6-1) with $S_n(\mathbb{K})$, the linear subspace of symmetric matrices in $M_n(\mathbb{K})$, and use the norm which $S_n(\mathbb{K})$ inherits from $M_n(\mathbb{K})$.

In the case char $\mathbb{K} = 2$, and only in this case, we also introduce

$$Q'(X) := \{f' \mid f'(x) = \sum_{j=1}^{n} a_j^2 x_j^2,\ a_j \in \mathbb{K},\ \text{for } x = \sum_{j=1}^{n} a_j e_j\}. \qquad (6\text{-}2)$$

Thus $Q'(X)$ consists of the squares of all linear forms on X. Hence it is a subgroup of the additive group of $Q(X)$, independent of the basis chosen in X.

5.7 We can now formulate an extension of the lemmas in §§ 5.4, 5.5 to arbitrary dimension, as announced at the end of § 5.5. It consists of the following *neighbourhood lemma for* $Q(X)$.

LEMMA. *Let* $f^\circ \in Q(X)$ *be defined as follows, by means of a chosen basis* $X^\circ = \{e_1, \dots, e_n\}$ *of* X. *Let for* $x = \sum_{j=1}^{n} x_j e_j$

$$f^\circ(x) := \sum_{j=1}^{n} x_j^2, \text{ if char } \mathbb{K} \ne 2,$$

$$f^\circ(x) := x_1^2,\ n = 1,\ f^\circ(x) := x_1^2 + \sum_{j=1}^{n-1} x_{j+1}x_j,\ n > 1, \text{ if char } \mathbb{K} = 2.$$

Let U *be any neighbourhood of* 1_n *in* $GL_n(\mathbb{K})$. *Let* $\alpha \in GL_n(\mathbb{K})$ *and write* $f^\circ \circ \alpha$ *for* $x \longmapsto f^\circ(x\alpha)$, $x \in X$. *Further let*

$$V_U(f^\circ) := \{f^\circ \circ \alpha \mid \alpha \in U\}, \text{ if char } \mathbb{K} \ne 2,$$

$$V_U(f^\circ) := \{f^\circ \circ \alpha + f' \mid \alpha \in U,\ f' \in Q'(X)\}, \text{ if char } \mathbb{K} = 2 \text{ [cf. (6-2)]}.$$

Then the set $V_U(f^\circ)$ *is a neighbourhood of* f° *in* $Q(X)$.

REMARKS.

1. The choice of the basis X° in the lemma above is irrelevant: a different choice amounts to replacing f° by $f^\circ \circ \alpha_1$, $\alpha_1 \in GL_n(\mathbb{K})$.

2. The quadratic form f° used in the case char $\mathbb{K} \ne 2$ cannot be used

when char $\mathbb{K} = 2$ and $n > 1$: for this f° the set $V_U(f^\circ)$ _is not a neighbourhood of_ f° _in_ $Q(X)$, _if_ char $\mathbb{K} = 2$, but an n-dimensional subspace, as is readily seen. It was only after the examination of several small values of n that a quadratic form suitable for the case of char $\mathbb{K} = 2$ was found.

3. For $n = 1$, the lemma above follows from the two lemmas in §§ 5.4, 5.5. We shall prove in § 6 _precise_ extensions of these two lemmas to higher dimensions which will imply the neighbourhood lemma. The reader may wish to go on immediately to §§ 7 and 8 in order to see the application of the neighbourhood lemma in the context of metaplectic groups.

§6. *Properties of certain quadratic forms*

We prove here the neighbourhood lemma for Q(X) [§ 5.7]. In fact, we shall prove two propositions, for the cases char $\mathbb{K} \neq 2$ and char $\mathbb{K} = 2$, respectively, which are rather more precise and entail this lemma. They are extensions of the lemmas in §§ 5.4, 5.5 to arbitrary dimension. The proofs require some careful analysis; the reader may wish to omit details in a first reading.

In § 6, the letters σ, τ will denote matrices.

6.1 Let $T_n(\mathbb{K})$ be the linear subspace (in fact, subalgebra) of $M_n(\mathbb{K})$ formed by the lower triangular matrices $\tau = (t_{ij})$, $t_{ij} \in \mathbb{K}$, $t_{ij} = 0$ for $j > i$. Let $S_n(\mathbb{K})$ be the linear subspace of all symmetric matrices $\sigma = {}^t\sigma$ in $M_n(\mathbb{K})$. We put $T_n(\mathbb{K})^* := T_n(\mathbb{K}) \cap GL_n(\mathbb{K})$, a group. In $\mathbb{K}^n$ we have introduced a certain norm, and likewise in $M_n(\mathbb{K})$ [§ 5.2].

If $\mathbb{K}$ is ultrametric, we define the following balls and other neighbourhoods, with q as in § 5.1:

$$B_N(1_n; T_n) := \{\tau \mid \tau \in T_n(\mathbb{K}),\ \|\tau - 1_n\| \leq q^{-N}\},\quad N = 1, 2, \ldots .$$

These are, in fact, subgroups of $T_n(\mathbb{K})^*$. We put, with the notation of § 5.1,

$$U_n(1_n; S_n) := \left\{ \sigma \mid \sigma = (s_{ij}) \in S_n(\mathbb{K}),\ \begin{array}{l} s_{ij} \in \mathfrak{p}^N, i \neq j, \\ s_{ii} \in 1 + \mathfrak{p}^{N+\mathrm{ord}(2)} \end{array} \right\}.$$

We also put

$$B_N(\mathbb{K}^n) := \{x \mid x \in \mathbb{K}^n,\ \|x\| \leq q^{-N}\}.$$

If $\mathbb{K}$ is $\mathbb{R}$ or $\mathbb{C}$, we let

$$B(1_n, \varepsilon; T_n) := \{\tau \mid \tau \in T_n(\mathbb{K}),\ \|\tau - 1_n\| < \varepsilon\},\quad 0 < \varepsilon < 1,$$

$$B(1_n, \varepsilon; S_n) := \{\sigma \mid \sigma \in S_n(\mathbb{K}),\ \|\sigma - 1_n\| < \varepsilon\},\quad 0 < \varepsilon < 1.$$

If char $\mathbb{K} \neq 2$, we define $f^\circ \in Q_n(\mathbb{K})$ by

$$f^\circ(x) := x.{}^tx,\quad x \in \mathbb{K}^n, \tag{1-1}$$

as in the neighbourhood lemma 5.7. We then introduce the following neighbourhoods of f°:

If char $\mathbb{K} \neq 2$ and $\mathbb{K}$ is ultrametric, we put

$$U_N(f^\circ; Q_n) := \{f \mid f(x) = x\sigma{}^tx,\ \sigma \in U_N(1_n; S_n)\},\quad N = 1, 2, \ldots ,$$

and if $\mathbb{K}$ is $\mathbb{R}$ or $\mathbb{C}$ we put

$$B(f^\circ,\varepsilon;Q_n) := \{f \mid f(x) = x\sigma{}^t x,\ \sigma\in B(1_n,\varepsilon;S_n)\},\quad 0 < \varepsilon < 1.$$

We further put for $\mathbb{K} = \mathbb{R}$ or $\mathbb{C}$ and $0 < \varepsilon < 1$

$$B(\varepsilon;\mathbb{K}^n) := \{x \mid x\in\mathbb{K}^n,\ \|x\| < \varepsilon\}, \qquad (1\text{-}2)$$

$$B(1_n,\varepsilon;\mathbb{K}^n) := \{x \mid x\in\mathbb{K}^n,\ \|x - 1_n\| < \varepsilon\} = 1_n + B(\varepsilon;\mathbb{K}^n). \qquad (1\text{-}3)$$

[For $n = 1$, these nds. were already introduced im § 5 (3-1).]

With these notations, we can now state a proposition which implies the case char $\mathbb{K} \neq 2$ of the neighbourhood lemma, § 5.7.

<u>6.2</u> PROPOSITION . *Let $\mathbb{K}$ be a local field*, char $\mathbb{K} \neq 2$. *Then, with the notations of* § 6.1, *we have*:

(i) *If $\mathbb{K}$ is ultrametric, the mapping $\tau \longmapsto \tau\cdot{}^t\tau$ of $T_n(\mathbb{K})$ into $S_n(\mathbb{K})$ yields, for $N > \mathrm{ord}(2)$, a homeomorphism of $B_N(1_n;T_n)$ - which is a compact, open subgroup of $T_n(\mathbb{K})^*$ - onto $U_N(1_n;S_n)$, a compact, open subgroup of the additive group $S_n(\mathbb{K})$. In other words, the mapping of $T_n(\mathbb{K})$ into $Q_n(\mathbb{K})$ defined by $\tau \longmapsto f^\circ\circ\tau$, with f° as in* (1-1), *yields a homeomorphism of $B_N(1_n;T_n)$ onto $U_N(f^\circ;Q_n)$ if $N > \mathrm{ord}(2)$*. [Cf. § 5 (1-3) and part (i) of the lemma in § 5.4.]

(ii) *If $\mathbb{K}$ is $\mathbb{R}$ or $\mathbb{C}$, then, for $0 < \varepsilon \le 1/2$, the mapping $\tau \longmapsto \tau\cdot{}^t\tau$ of $T_n(\mathbb{K})$ into $S_n(\mathbb{K})$ maps the ball $B(1_n,\varepsilon;T_n)$ - which lies, in fact, in $T_n(\mathbb{K})^*$ - injectively into $S_n(\mathbb{K})$*, i.e. *$\tau \longmapsto f^\circ\circ\tau$ is an injective mapping of $B(1_n,\varepsilon;T_n)$ into $Q_n(\mathbb{K})$; moreover, the image of $B(1_n,\varepsilon;T_n)$ contains the ball $B(f^\circ,\varepsilon/2;Q_n)$*. [Cf. part (ii) of the lemma in § 5.4.]

<u>Proof of part</u> (i). For $n = 1$, this is part (i) of the lemma in § 5.4. Suppose (i) holds and apply induction: consider matrices

$$\sigma_{n+1}\in S_{n+1}(\mathbb{K}),\quad \tau_{n+1}\in T_{n+1}(\mathbb{K})^*. \qquad (2\text{-}1)$$

Then we have

$$\sigma_{n+1} = \begin{pmatrix} \sigma_n & {}^t x' \\ x' & \xi' \end{pmatrix}, \qquad (2\text{-}2a)$$

$$\sigma_n\in S_n(\mathbb{K}),\ x'\in\mathbb{K}^n,\ \xi'\in\mathbb{K}, \qquad (2\text{-}2a')$$

$$\tau_{n+1} = \begin{pmatrix} \tau_n & {}^t 0 \\ x & \xi \end{pmatrix}, \qquad (2\text{-}2b)$$

$$\tau_n\in T_n(\mathbb{K})^*,\ x\in\mathbb{K}^n,\ 0\in\mathbb{K}^n,\ \xi\in\mathbb{K}. \qquad (2\text{-}2b')$$

Conversely, if (2-2a'), (2-2b') hold and σ_{n+1}, τ_{n+1} are <u>defined</u> by (2-2a), (2-2b), then (2-1) holds. Moreover, the relation

$$\sigma_{n+1} = \tau_{n+1}\cdot{}^t\tau_{n+1} \qquad (2\text{-}3)$$

is equivalent to the three relations

$$\text{(a) } \sigma_n = \tau_n \cdot {}^t\tau_n, \quad \text{(b) } x' = x \cdot {}^t\tau_n, \quad \text{(c) } \xi' = \xi^2 + x \cdot {}^t x. \tag{2-4}$$

Now observe that

$$\sigma_{n+1} \in U_N(1_{n+1}; S_{n+1}) \Leftrightarrow \sigma_n \in U_N(1_n; S_n),\ x' \in B_N(\mathbb{K}^n),\ \xi' \in 1+\mathfrak{p}^{N+\mathrm{ord}(2)}, \tag{2-5a}$$

$$\tau_{n+1} \in B_N(1_{n+1}; T_{n+1}) \Leftrightarrow \tau_n \in B_N(1_n; T_n),\ x \in B_N(\mathbb{K}^n),\ \xi \in 1+\mathfrak{p}^N. \tag{2-5b}$$

By the induction hypothesis, (2-4a) gives a bijective mapping $\tau_n \longmapsto \sigma_n$ of $B_N(1_n; T_n)$ onto $U_N(1_n; S_n)$. Next (2-4b), for $\tau_n \in B_N(1_n; T_n)$, is a bijective mapping $x \longmapsto x'$ of $B_N(\mathbb{K}^n)$ onto itself (since for such τ_n $\|x \cdot {}^t\tau_n\| = \|x\|$). Finally, if $x \in B_N(\mathbb{K}^n)$, then $x \cdot {}^t x \in \mathfrak{p}^{2N} \subset \mathfrak{p}^{N+\mathrm{ord}(2)}$, hence (2-4c) gives a bijective mapping $\xi \longmapsto \xi'$ of $1+\mathfrak{p}^N$ onto $1+\mathfrak{p}^{N+\mathrm{ord}(2)}$, by part (i) of the lemma in § 5.4. This yields part (i) of the proposition.

Proof of part (ii). For $n = 1$, the statement is a weaker formulation of part (ii) of the lemma in § 5.4. We again apply induction, based on the formulae (2-1) - (2-4). We put

$$\varepsilon' := n\varepsilon/(n+1), \quad 0 < \varepsilon \le 1/2,$$

so that

$$\varepsilon'/n = \varepsilon/(n+1) \text{ and } 0 < \varepsilon' < \varepsilon.$$

In view of the definition of the norms in $\mathbb{K}^n$, $M_n(\mathbb{K})$ and $M_{n+1}(\mathbb{K})$ for the case of $\mathbb{R}$ or $\mathbb{C}$ [§ 5.2], we have in the context of formulae (2-1) and (2-2a,a'), (2-2b,b') the following analogues of (2-5a,b):

$$\|\sigma_{n+1} - 1_{n+1}\| < \varepsilon \Leftrightarrow \|\sigma_n - 1_n\| < \varepsilon',\ \|x'\| < \varepsilon',\ |\xi'-1| < \varepsilon'/n,$$

$$\|\tau_{n+1} - 1_{n+1}\| < \varepsilon \Leftrightarrow \|\tau_n - 1_n\| < \varepsilon',\ \|x\| < \varepsilon',\ |\xi-1| < \varepsilon'/n.$$

By the hypothesis of the induction, the mapping $\tau_n \longmapsto \sigma_n$ given by (2-4a) maps $B(1_n, \varepsilon'; T_n)$ injectively into $S_n(\mathbb{K})$, and the image of $B(1_n, \varepsilon'; T_n)$ contains $B(1_n, \varepsilon'/2; S_n)$. Consider (2-4b) for τ_n in $B(1_n, \varepsilon'; T_n)$: here $\varepsilon' < 1/2$, so τ_n^{-1} exists and $\|\tau_n^{-1} - 1_n\| < 1$ [cf. § 5 (2-11)]. Hence for such a τ_n (2-4b) is an injective mapping $x \longmapsto x'$ of $B(\varepsilon'; \mathbb{K}^n)$ [cf. (1-2)] into $\mathbb{K}^n$, and the image of $B(\varepsilon'; \mathbb{K}^n)$ contains $B(\varepsilon'/2; \mathbb{K}^n)$, since

$$\|x' \cdot {}^t\tau^{-1}\| \le \|x' \cdot ({}^t\tau^{-1} - 1_n)\| + \|x'\| \le 2\|x'\| < \varepsilon' \text{ for } x' \in B(\varepsilon'/2; \mathbb{K}^n).$$

Finally, we note that

$$\text{if } x \in B(\varepsilon'; \mathbb{K}^n), \text{ then } |x \cdot {}^t x| < n \cdot (\varepsilon'/n)^2 < (1/2) \cdot (\varepsilon'/n).$$

Thus for such an x (2-4c) is an injective mapping $\xi \longmapsto \xi'$ of

$B(1,\varepsilon'/n;\mathbb{K})$ [notation as in (1-3) for n = 1] into $\mathbb{K}$, by § 5.4 (ii), and the image of $B(1,\varepsilon'/n;\mathbb{K}^n)$ contains $B(1,\varepsilon'/2n;\mathbb{K})$: for $x \in B(\varepsilon';\mathbb{K}^n)$ and $\xi' \in \mathbb{K}$ such that $|\xi'-1| < \varepsilon'/(2n)$ we have $|\xi' - x\cdot{}^t x - 1| < \varepsilon'/n$, so that, again by § 5.4 (ii), there is a (unique) ξ in $B(1,\varepsilon'/n;\mathbb{K})$ satisfying (2-4c).

Part (ii) of the proposition is now proved. This completes the proof of the proposition.

6.3 When char $\mathbb{K} = 2$, the situation is somewhat different: we have to work directly with the quadratic forms, i.e. we can no longer work with symmetric matrices. We write $f \in Q_n(\mathbb{K})$ as

$$f(x) = \sum_{1\le j\le k\le n} c_{kj}x_kx_j,\quad c_{kj}\in\mathbb{K},\quad x = (x_1,\dots,x_n)\in\mathbb{K}^n.$$

We recall that $Q_n(\mathbb{K})$ is isomorphic to $\mathbb{K}^{n(n+1)/2}$ as a Banach space over $\mathbb{K}$ [§ 5.6].

The result to be proved concerns the form

$$f^\circ = f_n^\circ:\ \begin{cases} f_n^\circ(x) := x_1^2, & n = 1,\\ f_n^\circ(x) := x_1^2 + \sum_{k=1}^{n-1} x_{k+1}x_k, & n > 1,\end{cases}\qquad x = (x_1,\dots,x_n)\in\mathbb{K}^n,$$

appearing in the neighbourhood lemma [§ 5.7] for the case char $\mathbb{K} = 2$; but we shall prove a more precise result than stated there. First we introduce various neighbourhoods and some notation.

Let $V_N(M_n)$ consist of all $\alpha = (a_{kj}) \in M_n(\mathbb{K})$ such that

$a_{kj} \in \mathfrak{p}^{2N}$ for k odd, j even,

$a_{kj} \in \mathfrak{p}^{N}$ for all other pairs (k, j), $1 \le k,j \le n$.

Next put $V_N(1_n;T_n^*) := (1_n+V_N(M_n))\cap T_n(\mathbb{K})^*$. Thus $V_N(1_n;T_n^*)$ consists of all matrices $\tau = (t_{kj})$ such that

$t_{kk} \in 1+\mathfrak{p}^N$, $1\le k\le n$; $t_{kj}\in\mathfrak{p}^{2N}$, k odd, j even, $1\le j<k\le n$,

$t_{kj}\in\mathfrak{p}^N$ for all other pairs (k, j), $1\le j<k\le n$, $t_{kj} = 0$, $n\ge j>k$.

Now let $V_N(f^\circ;Q_n)$ consist of all f, $f(x) = \sum_{1\le j\le k\le n} d_{kj}x_kx_j$, such that

$d_{11} \in 1+\mathfrak{p}^{2N}$

and for $2 \le k \le n$, $1 \le j \le k$,

$d_{k,k-1} \in 1+\mathfrak{p}^N$, $d_{kj}\in\mathfrak{p}^{2N}$ if k, j are both odd,

$d_{kj}\in\mathfrak{p}^N$ for all other pairs (k, j).

We put

$B_N(\mathbb{K}^n) := \{x \mid x = (x_1,\dots,x_n)\in\mathbb{K}^n,\ x_j\in\mathfrak{p}^N,\ 1\le j\le n\} =$

$$= \{x \mid x \in \mathbb{K}^n,\ \|x\| \le q^{-N}\}.$$

All these neighbourhoods are compact and open.

Finally, we let

$n' := [(n+1)/2]$, the largest integer contained in $(n+1)/2$,

and define for $c \in \mathbb{K}^{n'}$, i.e. $c = (c_k)_{1\le k\le n'}$, the quadratic form g_c° in $Q_n(\mathbb{K})$ by

$$g_c^\circ(x) := \sum_{k=1}^{n'} c_k{}^2 x_{2k-1}^{\;2},\quad x = (x_1,\dots,x_n) \in \mathbb{K}^n.$$

We can now state the result to be proved.

6.4 PROPOSITION. *Let $\mathbb{K}$ be a local field of characteristic 2; thus $\mathbb{K} \cong \mathbb{F}_q((T))$, $q = 2^r$. Then, with the notations in § 6.3, the mapping*

$$(\tau, c) \longmapsto f^\circ \circ \tau + T\cdot g_c^\circ$$

gives a continuous bijection of $V_N(1_n; T_n^*)\times B_N(\mathbb{K}^{n'})$ onto $V_N(f^\circ; Q_n)$, hence a homeomorphism.

This proposition is clearly an extension of the lemma in § 5.5 to higher dimensions; moreover, it entails the neighbourhood lemma in § 5.7 when the characteristic of $\mathbb{K}$ is 2.

The proof will be given in four parts. For general $n \ge 1$ various possible cases have to be considered. This makes the proof somewhat intricate. But its structure already becomes apparent for small n, and it will be instructive to specialize the proof below for n = 2, 3, 4, say, in order to see what happens in the general case.

(i) Let $f^\circ = f_n^\circ \in Q_n(\mathbb{K})$ be as stated in the proposition. *The first step of the proof is the determination of the coefficients of the quadratic form $f_n^\circ \circ \tau$ for general $\tau = (t_{ij})$ in $T_n(\mathbb{K})^*$.* We put

$$x = y\tau, \text{ or } x_j = \sum_{k=j}^{n} t_{kj} y_k,\quad 1 \le j \le n,\ y \text{ in } \mathbb{K}^n. \tag{4-1}$$

Then we write

$$f_n^\circ(x) = f_n^\circ(y\tau) = \sum_{k=1}^{n} \sum_{j=1}^{k} c_{kj} y_k y_j. \tag{4-2}$$

To determine the coefficients c_{kj} we observe that in (4-1) y_k occurs, for each k, only in x_j, $1 \le j \le k$. Thus in (4-2) the term with $y_1^{\,2}$ occurs only in $x_1^{\,2}$, so

$$c_{11} = (t_{11})^2 \tag{C_1'}$$

(we use here a special notation to denote this equation; it will appear later as part of a general system).

For $n > 1$ consider the product $y_k y_1$, with $1 < k \le n$: in $f_n^\circ(x)$ this product occurs once in the term $x_2 x_1$ (in x_1^2 it does not appear, since char K = 2 !). Hence we have

$$c_{k1} = t_{k2} t_{11}, \quad 1 < k \le n, \text{ for } n > 1. \qquad (C^{n,1})$$

Next consider, for $n > 2$, the product $y_k y_j$, $1 < j < k \le n$: this occurs in $f_n^\circ(x)$ twice in each term $x_{i+1} x_i$, $1 \le i < j$, once in the term $x_{j+1} x_j$: thus

$$\left.\begin{array}{l} c_{kj} = \sum_{i=1}^{j-1} (t_{k,i+1} \cdot t_{ji} + t_{ki} \cdot t_{j,i+1}) + t_{k,j+1} \cdot t_{jj}, \\ \qquad 1 < j < k \le n, \text{ for } n > 2. \end{array}\right\} \qquad (C^{n,2})$$

Lastly consider y_k^2, $1 \le k \le n$: this occurs in $f_n^\circ(x)$ once in x_1^2 and, when $n > 1$, also once in each term $x_{j+1} x_j$, $1 \le j < k$; hence

$$c_{kk} = (t_{k1})^2 + \sum_{j=1}^{k-1} t_{k,j+1} t_{kj}, \quad 1 < k \le n, \text{ for } n > 1. \qquad (C^{n,3})$$

The formulae $(C^{n,r})$, $r = 1, 2, 3$, together with (C_1') determine the $n(n+1)/2$ coefficients of $f_n^\circ(y\tau)$ for $n > 2$, and similarly for $n = 1, 2$.

REMARK. The above formulae show that <u>the coefficient</u> c_{kj}, $1 \le j \le k \le n$, <u>depends on</u> $t_{k'j'}$ <u>with</u> $j' \le k' \le k$, <u>but not on</u> $t_{k'j'}$, $k < k' \le n$, $j' \le k'$; in particular, c_{kj} <u>is the same for all forms</u> f_m° <u>with</u> $m \ge k$. Thus, if we pass from n to $n+1$, the formula (C_1') remains and the systems $(C^{n,r})$, $r = 1,2,3$, form part of the systems $(C^{n+1,r})$, respectively: we obtain the latter simply by adding new formulae, corresponding to $k = n+1$, to the former: one to $(C^{n,1})$, $n-1$ to $(C^{n,2})$, one to $(C^{n,3})$, for $n > 2$, and similarly for $n = 1, 2$. This fact makes it possible to use induction in proving the proposition, as we shall see in part (iv) of the proof.

(ii) <u>The next task is to simplify the systems</u> $(C^{n,1})$ <u>and</u> $(C^{n,2})$. For this purpose we introduce elements $b_{k,j}$ of $\mathbb{K}$, for $k \ge 2$, by the relations

$$\left.\begin{array}{l} b_{kj} = t_{k,j-1} + t_{k,j+1} \text{ for } 1 < j < k \le n, \ n \ge 3, \\ b_{k1} = t_{k2} \text{ for } 2 \le k \le n, \ n \ge 2. \end{array}\right\} \qquad (B_n)$$

Since char $\mathbb{K} = 2$, we readily see that (B_n) <u>implies</u>

$$\left.\begin{array}{l} t_{kh} = \sum_{\substack{j \text{ odd} \\ 1 \le j < h}} b_{kj}, \text{ if } h \text{ is even, } h \ge 2, \ h \le k \le n, \ n \ge 2, \\ t_{kh} = t_{k1} + \sum_{\substack{j \text{ even} \\ 2 \le j < h}} b_{kj}, \text{ if } h \text{ is odd, } h \ge 3, \ h \le k \le n, \ n \ge 3. \end{array}\right\} \qquad (B_n')$$

It is also readily seen that, conversely, (B_n') <u>implies</u> (B_n). Thus (B_n)

<u>and</u> (B'_n) <u>are equivalent</u>.

The systems $(C^{n,1})$ and $(C^{n,2})$ can now be combined: we have

$$c_{kj} = \sum_{i=1}^{j} b_{ki}t_{ji}\ , \quad 1 \le j < k \le n, \text{ for } n \ge 2. \qquad (C_n)$$

<u>It will also be useful</u>, for part (iv) of the proof, <u>to modify the system</u> $(C^{n,3})$ by introducing the elements b_{kj} there as well: this can be done by combining the terms in the sum on the right in $(C^{n,3})$ in pairs. If $k \ge 2$ is even, the last term $t_{kk}\cdot t_{k,k-1}$ is then left over and, if $k \ge 4$, we can transform it by applying (B'_n) with $h = k-1$. The system $(C^{n,3})$ can now be combined with formula (C'_1) and can be written as follows:

$$\left.\begin{aligned} c_{kk} &= (t_{k1})^2 + s_k\ , \text{ for } \underline{\text{odd}}\ k,\ 1 \le k \le n, \\ c_{kk} &= (t_{k1})^2 + t_{kk}\cdot t_{k1} + s_k + t_{kk}\cdot s'_k \text{ for } \underline{\text{even}}\ k,\ 2 \le k \le n, \end{aligned}\right\} \quad (C'_n)$$

where s_k, s'_k are defined by

$$\left.\begin{aligned} s_k &:= \sum_{j}^{\substack{j \text{ even} \\ 2\le j<k}} b_{kj}t_{kj} \text{ for } 3 \le k \le n;\ s_1 := 0,\ s_2 := 0, \\ s'_k &:= \sum_{j}^{\substack{j \text{ even} \\ 2\le j<k}} b_{kj} \quad \text{for } \underline{\text{even}}\ k,\ 2 < k \le n;\ s'_2 := 0. \end{aligned}\right\} \quad (4\text{-}3)$$

(s'_k is defined for even k only!) Again, the systems (C_n), (C'_n) are part of the systems (C_{n+1}), (C'_{n+1}) for $n \ge 2$, $n \ge 1$, respectively, in accordance with part (i) of the proof.

(iii) After these algebraic preliminaries, we can now show: <u>if</u> τ <u>is in</u> $V_N(1_n; T^*_n)$ <u>and</u> c <u>is in</u> $B_N(\mathbb{K}^{n'})$, <u>then</u> $f^\circ\circ\tau + T\cdot g^\circ_c$ <u>lies in</u> $V_N(f^\circ; Q_n)$.

The coefficients c_{kj} of $f^\circ\circ\tau$ are given by (C'_1), $(C^{n,r})$, $r = 1,2,3$. By (C'_1) c_{11} lies in $1+\mathfrak{p}^{2N}$ [since char $\mathbb{K} = 2$], and using (C_n) instead of $(C^{n,1})$, $(C^{n,2})$, we see: c_{kj} lies (at the least) in $\mathfrak{p}^N$ if $1\le j<k-1$ [for then, in (C_n), b_{ki} is in $\mathfrak{p}^N$, at the least!], while $c_{k,k-1}$ is in $1+\mathfrak{p}^N$ for $k > 1$ [for, $b_{k,k-1}$ and $t_{k-1,k-1}$ are in $1+\mathfrak{p}^N$]. Moreover, by assumption, t_{kh} is in $\mathfrak{p}^{2N}$ for odd k if h is even and $2\le h\le k-1$, so [cf. (B_n)] b_{kj} is in $\mathfrak{p}^{2N}$ when k and j are odd and $1\le j\le k-2$, hence c_{kj} is in $\mathfrak{p}^{2N}$ if k, j are odd, $j < k$. From $(C^{n,3})$ it follows that c_{kk} is in $\mathfrak{p}^N$, $1 < k \le n$; moreover, when $k > 1$ is odd, then c_{kk} even lies in $\mathfrak{p}^{2N}$ [by assumption, $t_{k,k-1}$ is in $\mathfrak{p}^{2N}$ if $k > 1$ is odd!].

Thus $f^\circ\circ\tau$ is in $V_N(f^\circ; Q_n)$. Since c is as stated, the coefficients of g°_c are in $\mathfrak{p}^{2N}$. Hence (iii) is proved.

(iv) In the final part of the proof we show the converse of (iii): <u>given</u> f <u>in</u> $V_N(f^\circ; Q_n)$, <u>there is a unique matrix</u> τ <u>in</u> $V_N(1_n, T^*_n)$, <u>and a</u>

<u>unique vector</u> c <u>in</u> $B_N(\mathbb{K}^{n'})$, <u>such that</u>

$$f = f^\circ \circ \tau + T\cdot g_c^\circ \, . \tag{4-4}$$

The proof is by induction. Let us write, for $n \geq 1$,

$$f(y) = f_n(y) = \sum_{1\leq j\leq k\leq n} d_{kj} y_k y_j \, . \tag{4-5}$$

For <u>$n = 1$</u> we have $f^\circ(x) = x_1^{\,2}$, and the statement is simply part of the lemma in § 5.5, stated in a slightly different way: we can write

$$d_{11} = (t_{11})^2 + T\cdot c_1^{\,2}, \tag{D_1'}$$

with t_{11}, c_1 uniquely determined by d_{11}, and if f is in $V_N(f^\circ;Q_1)$, i.e. $d_{11} \in 1+\mathfrak{p}^{2N}$, then t_{11} is in $1+\mathfrak{p}^N =: V_N(1;T_1^*)$, c_1 in $\mathfrak{p}^N =: B_N(\mathbb{K})$.

For <u>$n = 2$</u> we have $f^\circ(x) = x_1^{\,2} + x_1x_2$ and hence besides equation (D_1') also

$$d_{21} = t_{11}t_{22} \, , \tag{4-6}$$

$$d_{22} = (t_{21})^2 + t_{22}t_{21} \, . \tag{4-7}$$

From (D_1') we obtain unique elements $t_{11} \in 1+\mathfrak{p}^N$ and $c_1 \in \mathfrak{p}^N$, as before. We assert that then (4-6), (4-7) have unique solutions $t_{22} \in 1+\mathfrak{p}^N$ and $t_{21} \in \mathfrak{p}^N$: indeed, d_{21} is in $1+\mathfrak{p}^N$, d_{22} in $\mathfrak{p}^N$, by the definition of $V_N(f^\circ;Q_2)$; for t_{22} the assertion is thus obvious, since $1+\mathfrak{p}^N$ is a multiplicative group, for t_{21} it then follows from part (i) of the lemma in § 5.3.

Thus the proposition holds for n = 1, 2, in view of (iii).

Let us now consider the <u>general case</u> $n \geq 2$. We assume that the assertion at the beginning of part (iv) holds for $Q_n(\mathbb{K})$ - which in view of (iii) amounts to the validity of the proposition for $Q_n(\mathbb{K})$ - and wish to show that it also holds for $Q_{n+1}(\mathbb{K})$.

The relation (4-4) means, in terms of the coefficients of the quadratic form (4-5), the matrix $\tau \in T_n(\mathbb{K})^*$, and the vector $c \in \mathbb{K}^{n'}$, that, with the elements b_{kj} defined by (B_n), and s_k, s_k' defined by (4-3), the following formulae hold [cf. (C_n), (C_n')]:

$$d_{kj} = \sum_{i=1}^{j} b_{ki}t_{ji} \, , \; 1 \leq j < k \leq n, \text{ for } n > 1,$$

and for $1 \leq k \leq n$

$$d_{kk} = (t_{k1})^2 + s_k + T\cdot(c_{(k+1)/2})^2 \text{ for \underline{odd} } k,$$

$$d_{kk} = (t_{k1})^2 + t_{kk}t_{k1} + s_k + t_{kk}s_k' \text{ for \underline{even} } k,$$

or, since char $\mathbb{K} = 2$,

$$\left.\begin{aligned} d_{kk} + s_k &= (t_{k1})^2 + T\cdot(c_{(k+1)/2})^2 \text{ for \underline{odd} } k,\\ d_{kk} + s_k + t_{kk}s'_k &= (t_{k1})^2 + t_{kk}t_{k1} \text{ for \underline{even} } k, \end{aligned}\right\} 1 \le k \le n. \qquad (D'_n)$$

Notice that for $n = 2$ (D'_n) consists of (D'_1) together with (4-7), and (D_2) coincides with (4-6), since $s_1 = s_2 = 0$, $s'_2 = 0$. Quite generally, the systems (D_n), (D'_n), $n \ge 2$, are contained in (D_{n+1}), (D'_{n+1}), respectively, i.e. we obtain (D_{n+1}), (D'_{n+1}) from (D_n), (D'_n), respectively, by simply adding formulae of the same type, n to (D_n) and $n+1$ to (D'_n), corresponding to $k = n+1$. This simple relationship between (D_n) and (D_{n+1}), and between (D'_n) and (D'_{n+1}), is essentially that observed in part (i) of the proof, and makes induction possible.

The systems (D_n), (D'_n) are related to (4-4) and (4-5) as follows. Let τ be <u>any</u> matrix in $T_n(\mathbb{K})^*$ and c <u>any</u> vector in $\mathbb{K}^{n'}$, and let f be defined by (4-4). Then the relations (D_n), (D'_n) hold for the coefficients in (4-4). Moreover, the induction hypothesis says that conversely, if $f \in V_N(f^\circ;Q_n)$ is given by (4-5), then there is a unique matrix $\tau \in V_N(1_n,T^*_n)$, and a unique vector $c \in B_N(\mathbb{K}^{n'})$, satisfying (D_n), (D'_n), i.e. such that (4-4) holds.

To apply induction, let

$$\tau = (t_{kj}) \in V_N(1_n;Q_n), \quad c = (c_1,\dots,c_{n'}) \in B_N(\mathbb{K}^{n'})$$

be solutions of the systems (D_n), $(D_{n'})$, assumed to exist and to be unique. This is so for $n = 1$ (then there is only (D'_1)) and for $n = 2$, as shown above. Put, for shortness, $m := n+1$; <u>in considering (D_m), (D'_m), we have to distinguish two cases, according as m is odd or even</u>.

<u>Suppose that $m := n+1$ is odd</u>. Then, by the definition of $V_N(f^\circ_m;Q_m)$,

$$d_{mj} \in \begin{cases} \mathfrak{p}^{2N} \\ \mathfrak{p}^{N} \end{cases} \text{if } j \text{ is } \begin{cases} \text{odd} \\ \text{even} \end{cases} \text{and } j < m-1, \quad d_{m,m-1} \in 1+\mathfrak{p}^N. \qquad (4\text{-}8)$$

First, we can solve the equations

$$d_{mj} = \sum_{i=1}^{j} b_{mi}t_{ji}, \quad 1 \le j < m, \qquad (4\text{-}9)$$

uniquely for b_{mj}, step by step, starting with $j = 1$: by the induction hypothesis, we have $t_{jj} \in 1+\mathfrak{p}^N$, $1 \le j < m$, and $t_{ji} \in \mathfrak{p}^N$, $1 \le i < j < m$; this gives by (4-8), and since $1+\mathfrak{p}^N$ is a multiplicative group,

$$b_{mj} \in \begin{cases} \mathfrak{p}^{2N} \\ \mathfrak{p}^{N} \end{cases} \text{if } j \text{ is } \begin{cases} \text{odd} \\ \text{even} \end{cases} \text{and } j < m-1, \quad b_{m,m-1} \in 1+\mathfrak{p}^N. \qquad (4\text{-}10)$$

Then the first part of (B'_m), with m replacing n and $k = m$, determines unique elements t_{mh} for all <u>even</u> h, $2 \le h \le m-1$, and in view of (4-10) also

$$t_{mh} \in \mathfrak{p}^{2N} \text{ for all even } h,\ 2 \le h \le m-1. \tag{4-11}$$

Next, the first part of formula (4-3) gives unique elements s_m, if we replace there n by m and let k = m. Moreover, by (4-11), we have

$$s_m \in \mathfrak{p}^{2N}$$

[note in (4-3) that for k = m (instead of n) $t_{m,m-1}$ is in $\mathfrak{p}^{2N}$, while $b_{m,m-1}$ is in $1+\mathfrak{p}^N$, and for the terms in the sum s_m with $j < m-1$ each factor is at least in $\mathfrak{p}^N$].

Since d_{mm} is in $\mathfrak{p}^{2N}$ by the definition of $V_N(f^\circ_m; Q_m)$, elements t_{m1} in $\mathfrak{p}^N$ and $c_{(m+1)/2}$ in $\mathfrak{p}^N$ are uniquely determined by the relation [cf. (D'_n) and the lemma in § 5.5]

$$d_{mm} + s_m = (t_{m1})^2 + T\cdot(c_{(m+1)/2})^2.$$

Lastly, the second part of (B'_m), with m in place of n and k = m, determines unique elements t_{mh} for all <u>odd</u> h, $3 \le h \le m$, and in view of (4-10) t_{mh} is in $\mathfrak{p}^N$ if $h < m$, while t_{mm} is in $1+\mathfrak{p}^N$.

Thus we have shown that (D_{n+1}) and (D'_{n+1}) hold, with unique τ in $V_N(1_{n+1}; T^*_{n+1})$, and unique c in $B_N(\mathbb{K}^{(n+2)/2})$. Hence [cf. part (iii)] the proposition holds if n+1 is odd.

<u>Now suppose that</u> m := n+1 <u>is even</u>. Then, in the same way as before, (4-9) determines the elements b_{mj}, $1 \le j < m$, uniquely; moreover, by the definition of $V_N(f^\circ_m; Q_m)$, d_{mj} is in $\mathfrak{p}^N$ for $j < m-1$, $d_{m,m-1}$ is in $1+\mathfrak{p}^N$, so we have this time

$$b_{mj} \in \mathfrak{p}^N,\ j < m-1,\ b_{m,m-1} \in 1+\mathfrak{p}^N.$$

Again, by the first part of (B'_m), with m replacing n and k = m, unique elements t_{mh} are determined for all <u>even</u> $h \le m$, and

$$t_{mh} \in \mathfrak{p}^N \text{ for all even } h,\ 2 \le h < m,\ t_{mm} \in 1+\mathfrak{p}^N.$$

It follows that s_m and s'_m, uniquely determined by (4-3), are in $\mathfrak{p}^N$ (s_m lies in $\mathfrak{p}^{2N}$, in fact, but this is irrelevant here). Then the relation

$$d_{mm} + s_m + t_{mm}s'_m = (t_{m1})^2 + t_{mm}t_{m1}$$

determines a unique $t_{m1} \in \mathfrak{p}^N$, since here d_{mm} is in $\mathfrak{p}^N$, by the definition of $V_N(f^\circ_m; Q_m)$ [cf. § 5.3, Lemma, part (i)]. Finally, the elements

$$t_{mh} \in \mathfrak{p}^N,\ h \text{ odd},\ 3 \le h < m,$$

are uniquely determined by the second part of (B'_m), m replacing n.

Thus, in the same way as before, we obtain: the proposition holds in $Q_{n+1}(\mathbb{K})$ if n+1 is even. This completes the induction and concludes the proof of the proposition.

The propositions in § 6.2 and in § 6.4 yield, in particular, the neighbourhood lemma, § 5.7, as already noted.

In this § we have worked with $\mathbb{K}^n$ and its canonical basis. Is there, perhaps, a more invariant treatment? In the case of char $\mathbb{K} = 2$, can the quadratic form f° be characterized intrinsically ? Another question is whether the neighbourhood lemma in § 5.7 can be proved without any recourse to the far more precise propositions in § 6.2 and § 6.4. But these questions can only be stated here ... .

§7. *Weil operators for vector spaces over local fields*

7.1 Let us first consider a finite-dimensional vector space X over an arbitrary (commutative) field k. Let X^* be the algebraic dual of X; it has the same dimension, say n. We denote the value of the linear form $x^* \in X^*$ at $x \in X$ by $[x, x^*]$. By Aut(X/k) we mean the multiplicative group of all k-linear automorphisms of X. If we take a basis in X, we obtain an (algebraic) isomorphism of Aut(X/k) with $GL_n(k)$. Suppose k contains a proper subfield k_o such that $[k:k_o] < \infty$. Then X is also a finite-dimensional vector space over k_o, and Aut(X/k) is a proper subgroup of $\mathrm{Aut}(X/k_o)$.

Next consider $X \times X^*$. We put $z := (x, x^*) \in X \times X^*$ and write $z\sigma$ for the action of $\sigma \in \mathrm{Aut}(X \times X^*/k)$ on z; the notation here indicates again k-linearity. We can then write, similarly as in the case of groups [§ 4.1], with an obvious notation:

$$\sigma = \begin{pmatrix} \alpha & \beta \\ \gamma & \delta \end{pmatrix}, \quad \begin{matrix} \alpha \in \mathrm{Aut}(X/k), & \beta \in \mathrm{Mor}(X, X^*/k), \\ \gamma \in \mathrm{Mor}((X^*, X/k), & \delta \in \mathrm{Aut}(X^*/k). \end{matrix} \tag{1-1}$$

If b is any k-bilinear form on $X \times X$, then we have: $b(x,y) = [x, y\beta]$, $(x,y) \in X \times X$, for a unique $\beta \in \mathrm{Mor}(X, X^*/k)$. We define $\beta^* \in \mathrm{Mor}(X, X^*/k)$ by $[x, y\beta] = [y, x\beta^*]$. We further put

$$\left.\begin{aligned} & b_o(x, x^*) := [x, x^*], \\ & b_1(z_1, z_2) := [x_1, x_2^*] \text{ for } z_j = (x_j, x_j^*) \in X \times X^*, \ j = 1, 2. \end{aligned}\right\} \tag{1-2}$$

The (linear) symplectic group Sp(X/k) is defined as the group of all σ in $\mathrm{Aut}(X \times X^*/k)$ such that $b_1(z_1, z_2) - b_1(z_2, z_1)$ is invariant under $z_j \longmapsto z_j\sigma$, j = 1, 2, or $b_1(z_1\sigma, z_2\sigma) - b(z_1, z_2)$ is invariant under the interchange of z_1 and z_2.

The linear case and that of l.c.a. groups are entirely analogous, with a simple change of notation; cf. in particular § 1.4.

For the standard vector space k^n, with the canonical basis, we put

$$[x, x'] := x \cdot {}^t x', \quad x, x' \text{ in } k \tag{1-3}$$

(recall that vectors are written as rows). In this way $(k^n)^*$ appears as another copy of k^n.

Now consider $k^n \times k^n$. We have for (x,y), (x',y') in $k^n \times k^n$

$$[(x,y), (x',y')] = (x,y) \cdot {}^t(x',y') = x \cdot {}^t x' + y \cdot {}^t y'. \tag{1-4}$$

We introduce the canonical basis in each k^n. Then we can represent mappings by matrices,

$$\begin{pmatrix}\alpha & \beta\\ \gamma & \delta\end{pmatrix} \longmapsto \begin{pmatrix}A & B\\ C & D\end{pmatrix}, \tag{1-5}$$

i.e. we obtain an algebraic isomorphism of $\mathrm{Aut}(k^n\times k^n/k)$ onto $GL_{2n}(k)$. Relation (1-4) shows that to $\begin{pmatrix}\alpha & \beta\\ \gamma & \delta\end{pmatrix}^*$ corresponds the transposed matrix, i.e. $\begin{pmatrix}{}^tA & {}^tC\\ {}^tB & {}^tD\end{pmatrix}$. Thus we obtain in particular an isomorphism of $Sp(k^n/k)$ onto $Sp_{2n}(k)$.

7.2 Let us return to X and X*. If we choose a basis in X, we get a k-linear isomorphism, say ρ, of X onto k^n. Then the isomorphism ρ^{*-1} is the mapping of X* onto k^n obtained by choosing the dual basis in X*: for, the definition of ρ^*, analogous to that in § 1 (4-2), amounts to

$$[x,x^*] = (x\rho)^t(x^*\rho^{*-1}), \tag{2-1}$$

and if $(e_j)_{1\le j\le n}$, $(e_j^*)_{1\le j\le n}$ are the bases in question, then $(e_j\rho)_{1\le j\le n}$ is the canonical basis in k^n, by the definition of ρ, and by (2-1) $(e_j^*\rho^{*-1})_{1\le j\le n}$ is the dual basis, i.e. the canonical basis again. The isomorphism ρ also yields the following isomorphism of $\mathrm{Aut}(X\times X^*/k)$ onto $\mathrm{Aut}(k^n\times k^n/k)$:

$$\sigma = \begin{pmatrix}\alpha & \beta\\ \gamma & \delta\end{pmatrix} \longmapsto \sigma' := \begin{pmatrix}\rho^{-1} & 0\\ 0 & \rho^*\end{pmatrix}\begin{pmatrix}\alpha & \beta\\ \gamma & \delta\end{pmatrix}\begin{pmatrix}\rho & 0\\ 0 & \rho^{*-1}\end{pmatrix} =: \begin{pmatrix}\alpha' & \beta'\\ \gamma' & \delta'\end{pmatrix}, \tag{2-2a}$$

with

$$\alpha' := \rho^{-1}\alpha\rho,\ \beta' := \rho^{-1}\beta\rho^{*-1},\ \gamma' := \rho^*\gamma\rho,\ \delta' := \rho^*\delta\rho^{*-1}. \tag{2-2b}$$

This yields an isomorphism of $\mathrm{Aut}(X\times X^*/k)$ onto $GL_{2n}(k)$ [cf. (1-5)]; in particular we obtain an isomorphism of $Sp(X/k)$ onto $Sp(k^n/k)$ and hence onto $Sp_{2n}(k)$. In this context it is instructive to read Siegel's treatment of $Sp_{2n}(\mathbb{R})$ [24, Bd. II, Kap. 4, § 4, p. 150]; see also Freitag [8, Kap. 1].

The formulae (2-2b) show that under the mapping (2-2a) the special elements

$$\left.\begin{array}{l}\sigma_m(\alpha) := \begin{pmatrix}\alpha & 0\\ 0 & \alpha^{*-1}\end{pmatrix},\ \sigma_a(\beta) := \begin{pmatrix}1 & \beta\\ 0 & 1\end{pmatrix},\ \sigma_w(\gamma) := \begin{pmatrix}0 & -\gamma^{*-1}\\ \gamma & 0\end{pmatrix},\\ \alpha\in \mathrm{Aut}(X\times X^*/k),\ \beta\in \mathrm{Mor}(X,X^*/k),\ \beta^*=\beta,\ \gamma\in \mathrm{Is}(X^*,X/k),\end{array}\right\} \tag{2-3}$$

go over into the corresponding elements of $Sp(k^n/k)$:

$$\sigma_m(\alpha) \longmapsto \sigma_m'(\alpha'),\ \sigma_a(\beta) \longmapsto \sigma_a'(\beta'),\ \sigma_w(\gamma) \longmapsto \sigma_w'(\gamma') \tag{2-4}$$

with α', β', γ' defined by (2-2b) and $\sigma_m'(\alpha')$, $\sigma_a'(\beta')$, $\sigma_w'(\gamma')$ in $Sp(k^n/k)$ defined in analogy to (2-3).

We can now state <u>Witt's theorem in invariant form</u>. <u>Let</u> γ <u>be in</u> Is(X*,X/k). <u>The group</u> Sp(X/k) <u>is generated by the elements</u> $\sigma_w(\gamma)$, $\sigma_m(\alpha)$, $\alpha \in \mathrm{Aut}(X/k)$, <u>and</u> $\sigma_a(\beta)$, $\beta \in \mathrm{Mor}(X, X^*/k)$, $\beta^* = \beta$.

By (2-4) above, it is enough to take $X = k^n$, with the canonical basis; we may thus consider the matrix groups $\mathrm{Sp}_{2n}(k)$. The elements (2-3) of $\mathrm{Sp}(k^n/k)$ are then represented, respectively, by the matrices

$$\sigma_m(A) := \begin{pmatrix} A & 0 \\ 0 & {}^tA^{-1} \end{pmatrix}, \quad \sigma_a(B) := \begin{pmatrix} 1_n & B \\ 0_n & 1_n \end{pmatrix}, \quad \sigma_w(C) = \begin{pmatrix} 0 & -{}^tC^{-1} \\ C & 0 \end{pmatrix} = \sigma_w(1_n)\cdot\sigma_m(C)$$

of $\mathrm{Sp}_{2n}(k)$, where

$A \in GL_n(k)$, $B \in M_n(k)$ and ${}^tB = B$, $C \in GL_n(k)$.

These matrices, with C fixed, generate $\mathrm{Sp}_{2n}(k)$ - this is the classical result of Witt [28, Satz C, p.336, and the remark on p.336-337]; see also Freitag [8, A.5.4, p.326] for an interesting proof.

REMARK. In the theory of algebraic groups it is shown that, in fact, for any (commutative) field k, $\mathrm{Sp}_{2n}(k) = \Omega^2$, where

$$\Omega := \{\sigma_a(B_1)\cdot\sigma_w(C)\cdot\sigma_a(B_2) \mid B_j \in M_n(k),\ {}^tB_j = B_j,\ C \in GL_n(k)\},$$

i.e. Ω is simply the set of all $\begin{pmatrix} A & B \\ C & D \end{pmatrix}$ in $\mathrm{Sp}_{2n}(k)$ with $\det(C) \neq 0$ [cf. § 4 (3-2)]. A proof of this for $k = \mathbb{R}$ in Igusa's book [10, Ch.I, § 6, Lemma 6] holds, in fact, for any infinite field k. But we shall not need this; in the present context, Witt's theorem is sufficient and brings out the significance of the Weil operators (cf. below, Proposition 7.8).

<u>7.3</u> The foregoing §§ 7.1, 7.2 apply in particular if $k = \mathbb{K}$, a local field. Then X carries the product topology of $\mathbb{K}^n$, so X is locally compact. A fundamental theorem states that a finite-dimensional topological vector space over a local field has a unique topology (cf. Weil's book [26, Ch.I, § 2, Th.3 and Cor.1]). The mappings α, β, γ, δ in (1-1) are continuous, since they are $\mathbb{K}$-linear, so the morphisms (1-1) are all in the sense of <u>topological</u> vector spaces.

If $\mathbb{K}$ is a local field other than $\mathbb{R}$ or $\mathbb{Q}_p$, then $\mathbb{K}$ contains a subfield k_o such that $[\mathbb{K}:k_o] < \infty$. [for $\mathbb{K} = \mathbb{F}_q((T))$ one can take $\mathbb{F}_q((T^m))$ for any $m > 1$]. Thus, <u>if</u> $\mathbb{K} \neq \mathbb{R}$, $\mathbb{Q}_p$, <u>then</u> Sp(X/$\mathbb{K}$) <u>is a proper subgroup of</u> [Sp(X/k_o), hence of] Sp(X) in the sense of § 4.2 which only depends on the additive group of $\mathbb{K}$; <u>but for</u> $\mathbb{K} = \mathbb{R}$, $\mathbb{Q}_p$, <u>equality holds</u>: here additivity implies linearity over the rationals which are dense in $\mathbb{K}$, hence (by continuity) linearity over $\mathbb{K}$. Likewise for Aut(X/$\mathbb{K}$).

The isomorphism of Aut(X/$\mathbb{K}$) with $GL_n(\mathbb{K})$, defined by introducing a basis in X, holds in the sense of topological groups, Aut(X/$\mathbb{K}$) being provided with Braconnier's automorphism topology [§ 1.6] and $GL_n(\mathbb{K})$ with the norm topology [§ 5.2].

To prove this, let us identify Aut(X/$\mathbb{K}$) with $GL_n(\mathbb{K})$ by means of a basis of X and the canonical basis of $\mathbb{K}^n$. We take the operator norm $\| \ \|_o$ in $GL_n(\mathbb{K})$ [§ 5 (2-5)] and put $U_\varepsilon := \{\alpha \mid \alpha \in GL_n(\mathbb{K}),\ \|\alpha - 1_n\|_o < \varepsilon\}$, $0 < \varepsilon < 1$; this is a nd. basis at 1_n for the norm topology of $GL_n(\mathbb{K})$. We also have the nds. $\mathcal{U}(K,V)$ of 1_n in the automorphism topology of $GL_n(\mathbb{K})$ acting on $\mathbb{K}^n$. We want to show: the nd. bases U_ε $(0 < \varepsilon < 1)$ and $\mathcal{U}(K,V)$ ($K \subset \mathbb{K}^n$ compact, V a nd. of 0 in $\mathbb{K}^n$) are equivalent.

(i) Every $\mathcal{U}(K,V)$ contains some U_ε. This is shown as follows. First, we can take $\varepsilon' > 0$ such that $x\alpha' - x \in V$ for all $x \in K$, if $\alpha' \in M_n(\mathbb{K})$ and $\|\alpha' - 1_n\|_o < \varepsilon'$: for, if we put $B := \{x \mid x \in \mathbb{K}^n,\ \|x\| \le 1\}$, then we have $K \subset c_1 \cdot B$ and $c_2 \cdot B \subset V$ for suitable c_1, c_2 $(\neq 0)$ in $\mathbb{K}$, so we can put $\varepsilon' := |c_2|/|c_1|$. Then we take ε, $0 < \varepsilon < 1$, such that for all $\alpha \in M_n(\mathbb{K})$ with $\|\alpha - 1_n\|_o < \varepsilon$ we have $\|\alpha - 1_n\|_o < \varepsilon'$ and $\|\alpha^{-1} - 1_n\|_o < \varepsilon'$ [e.g. $\varepsilon := \varepsilon'/(1+\varepsilon')$ will do, by the analogue of § 5 (2-11) for the operator norm]. This shows indeed $U_\varepsilon \subset \mathcal{U}(K,V)$.

(ii) Every U_ε contains some $U(K,V)$. To show this, we take $K := B$, with B as in part (i), and $V := c \cdot B$ with $c \in \mathbb{K}$, $0 < |c| < \varepsilon$. If $\alpha \in GL_n(\mathbb{K})$ satisfies, for all $x \in \mathbb{K}^n$ with $\|x\| \le 1$, the inequalities $\|x(\alpha - 1_n)\| \le |c|$, $\|x(\alpha^{-1} - 1_n)\| \le |c|$, then $\|\alpha - 1_n\|_o \le |c| < \varepsilon$ (and $\|\alpha^{-1} - 1_n\|_o < \varepsilon$). Hence $\mathcal{U}(K,V) \subset U_\varepsilon$, and the proof is finished.

7.4 We list here the basic facts of harmonic analysis for the additive group of a local field $\mathbb{K}$. References are given at the end of this section.

A non-trivial character $\chi = \chi_{\mathbb{K}}$ of the additive group of $\mathbb{K}$ may be obtained as follows [cf. § 5.1]:

(I) If $\mathbb{K}$ is $\mathbb{R}$, we define $\chi_{\mathbb{R}}(x) := \underline{e}(x) := e^{2\pi i x}$, $x \in \mathbb{R}$. For $\mathbb{K} = \mathbb{C}$ we can put $\chi_{\mathbb{C}}(x) := \underline{e}(2 \cdot \mathrm{Re}(x))$, $x \in \mathbb{C}$ [$2 \cdot \mathrm{Re}(x)$ is the trace of x over $\mathbb{R}$].

(IIa) $\mathbb{Q}_p$ is the completion of $\mathbb{Q}$ for the p-adic absolute value. Let $\mathbb{Q}^{(p)}$ consist of all rationals of the form a/p^n, $n = 0,1,2,\ldots$, $a \in \mathbb{Z}$. Let Z_p' consist of all rationals r' with denominator not divisible by p, i.e. $|r'|_p \le 1$. Then $\mathbb{Q} = \mathbb{Q}^{(p)} + Z_p'$ and $\mathbb{Q}^{(p)} \cap Z_p' = \mathbb{Z}$. As usual, $\mathbb{Z}_p$ denotes the p-adic integers, i.e. all $x \in \mathbb{Q}_p$ with $|x|_p \le 1$; so $\mathbb{Z}_p$ is the closure of Z_p' in $\mathbb{Q}_p$. Now, if x_1, x_2 are elements of distinct cosets of the subgroup Z_p' in $\mathbb{Q}$, then we have $|x_1 - x_2| \ge p$, hence $\mathbb{Q}_p = \mathbb{Q}^{(p)} + \mathbb{Z}_p$; also, of course, $\mathbb{Q}^{(p)} \cap \mathbb{Z}_p = \mathbb{Z}$. So for any x in $\mathbb{Q}_p$ there is an $x^{(p)}$ in

$\mathbb{Q}^{(p)}$ such that $x-x^{(p)}$ lies in $\mathbb{Z}_p$. We can put $\chi_{\mathbb{Q}_p}(x) := \underline{e}(x^{(p)})$ which defines $\chi_{\mathbb{Q}_p}(x)$ uniquely. Then $\chi_{\mathbb{Q}_p}$ is a non-trivial character of (the additive group of) $\mathbb{Q}_p$; its kernel is $\mathbb{Z}_p$. For a finite algebraic extension $\mathbb{K}$ of $\mathbb{Q}_p$ we put $\chi_{\mathbb{K}}(x) := \chi_{\mathbb{Q}_p}(\mathrm{tr}(x))$, $x \in \mathbb{K}$ [trace of x over $\mathbb{Q}_p$].

(IIb) For $\mathbb{K} = \mathbb{F}_q((T))$, with $q = p^r$, let χ' be a non-trivial character of the additive group of $\mathbb{F}_q$ and put $\chi_{\mathbb{K}}(x) := \chi'(a_{-1})$ for $x = \sum_n a_n T^n$ in $\mathbb{K}$. Then $\chi_{\mathbb{K}}$ is a non-trivial character of $\mathbb{K}$. Its values are p-th roots of unity.

Every character of the additive group of $\mathbb{K}$ is of the form $x \longmapsto \chi_{\mathbb{K}}(cx)$, $c \in \mathbb{K}$. But this holds true in greater generality. Let X be a finite-dimensional vector space over $\mathbb{K}$, X* its algebraic dual. X and X* are topological vector spaces and are locally compact. <u>The additive group of X* is the dual, in the sense of harmonic analysis, of the additive group of X. The duality can be realized by taking a non-trivial character $\chi = \chi_{\mathbb{K}}$ of $\mathbb{K}$ and putting</u>

$$\langle x, x^* \rangle := \chi([x, x^*]), \quad x \in X, \ x^* \in X^*, \tag{4-1}$$

where $(x, x^*) \longmapsto [x, x^*]$ is the $\mathbb{K}$-bilinear form expressing the algebraic duality of V and V* ($[x, x^*]$ is the value of $x^* \in X^*$ at $x \in X$)

In the case of $\mathbb{K}^n$, with the canonical basis, we can take $(\mathbb{K}^n)^*$ as another copy of $\mathbb{K}^n$, with the canonical basis again, and use the bilinear form (1-3). We then take the <u>self-dual Haar measure on</u> $\mathbb{K}^n$, i.e. that Haar measure for which the constant in the Fourier inversion formula has the value 1 (this measure depends on the choice of the basic character χ in (4-1)). If dx is any Haar measure on $\mathbb{K}^n$, then the dual Haar measure on (the second copy of) $\mathbb{K}^n$ will be $c \cdot dx$, with a constant $c > 0$, and the self-dual Haar measure will thus be $\sqrt{c} \cdot dx$. But for our purposes we need not determine c explicitly.

REFERENCES. Weil [25, n° 24], [26, Ch. II, § 5], Cassels-Fröhlich [6, Ch. 15, § 2.2].

<u>7.5</u> For a finite-dimensional vector space X over a local field $\mathbb{K}$, the symplectic group $Sp(X/\mathbb{K})$, as defined in § 7.1, bears an obvious relation to the group Sp(X) which is attached to the additive group of $\mathbb{K}$ according to § 4.2:

$$Sp(X/\mathbb{K}) = Sp(X) \cap Aut(X \times X^*/\mathbb{K}). \tag{5-1}$$

[$Aut(X \times X^*/\mathbb{K})$ consists of the $\mathbb{K}$-<u>linear</u> automorphisms of $X \times X^*$ (§ 7.1).] For, here the F of § 4.2 has the form $F = \chi \circ b_1$ with b_1 as in (1-2). If σ is in Sp(X) and in $Aut(X \times X^*/\mathbb{K})$, then for z_1, z_2 in $X \times X^*$ we have

$$\chi(b_1(z_1\sigma,z_2\sigma) - b_1(z_2\sigma,z_1\sigma)) = \chi(b_1(z_1,z_2) - b_1(z_2,z_1)),$$

or, if we put $b'(z_1,z_2) := b_1(z_1,z_2) - b_1(z_2,z_1)$,

$$\chi(b'(z_1\sigma,z_2\sigma) - b'(z_1,z_2)) = 1 \text{ for all } z_1,\ z_2 \text{ in } X\times X^*.$$

Replacing here z_1 by $c\cdot z_1$, $c\in\mathbb{K}$, we have by the $\mathbb{K}$-linearity of σ

$$\chi(c\cdot[b'(z_1\sigma,z_2\sigma) - b'(z_1,z_2)]) = 1 \text{ for all } c\in\mathbb{K},$$

whence $b'(z_1\sigma,z_2\sigma) - b'(z_1,z_2) = 0$, so σ is in $Sp(X/\mathbb{K})$. The converse is trivial.

$Sp(X/\mathbb{K})$ <u>is a proper subgroup of</u> $Sp(X)$ <u>unless</u> $\mathbb{K}$ <u>is</u> $\mathbb{R}$ <u>or</u> $\mathbb{Q}_p$ (§ 7.3).

<u>7.6</u> From now on we shall only deal with the linear group $Sp(X/\mathbb{K})$. In accordance with (5-1), we shall thus consider only the following Weil operators on $L^2(X)$:

(i) $M(\alpha)$, $\alpha\in Aut(X/\mathbb{K})$,

(ii) W^γ, $\gamma\in Is(X^*,X/\mathbb{K})$ [$\mathbb{K}$-linear isomorphisms of X^* onto X],

(iii) $A(\psi)$, $\psi\in Ch_2(X/\mathbb{K})$,

where $Ch_2(X/\mathbb{K})$ is defined as the multiplicative group of all characters of the 2nd degree on X such that, with $\chi = \chi_{\mathbb{K}}$ [cf. § 7.4],

$$\psi(x+y) = \psi(x)\psi(y)\chi(b(x,y)),\quad x,\ y \text{ in } X, \tag{6-1}$$

b being a $\mathbb{K}$-<u>bilinear</u> form (thus continuous!) on $X\times X^*$, necessarily symmetric. Equivalently, we can write

$$\psi(x+y) = \psi(x)\psi(y)\chi([x,y\beta]), \text{ with } \beta\in Mor(X,X^*/\mathbb{K}),\ \beta^* = \beta.$$

Now consider $X = \mathbb{K}^n$, with the canonical basis, and the bilinear form (1-3) on $\mathbb{K}^n\times\mathbb{K}^n$. The elements α in (i) above may then be interpreted as matrices: $\alpha\in GL_n(\mathbb{K})$. In (ii) we can take the identity mapping of $(\mathbb{K}^n)^* = \mathbb{K}^n$ onto $\mathbb{K}^n$: the corresponding Weil operator is simply the ordinary Fourier transform $\mathcal{F}$, $\mathcal{F}\Phi = \hat{\Phi}$, in $L^2(\mathbb{K}^n)$, with the self-dual Haar measure so that the operator is unitary. The general Weil operator W^γ can then be written $\mathcal{F}\cdot M(\alpha)$, $\alpha\in GL_n(\mathbb{K})$. In (iii) above we write, in the case of $\mathbb{K}^n$, for the form b in (6-1): $b(x,y) = x\beta\,{}^ty$, with $\beta\in M_n(\mathbb{K})$, ${}^t\beta = \beta$.

<u>7.7</u> It is of some interest to describe the elements of $Ch_2(\mathbb{K}^n/\mathbb{K})$ explicitly, for the canonical basis. That is, given a symmetric matrix $\beta\in M_n(\mathbb{K})$, we want to construct in $Ch_2(\mathbb{K}^n/\mathbb{K})$ one particular solution $\psi = \psi_\beta$ of the functional equation

$$\psi(x+y) = \psi(x)\psi(y)\chi(x\beta\,{}^ty)\quad [\chi = \chi_{\mathbb{K}}]. \tag{7-1}$$

The general solution will then be

$$\psi(x) = \psi_\beta(x)\cdot\chi(x^t c),\quad c\in \mathbb{K}^n.$$

If char $\mathbb{K} \neq 2$, then a particular solution is obviously given by

$$\psi_\beta(x) := \chi(2^{-1}x\beta^t x). \tag{7-2}$$

But if char $\mathbb{K} = 2$, i.e. if $\mathbb{K} = \mathbb{F}_q((T))$, $q = 2^r$, another method of solution is required. We know that a ψ satisfying (7-1) exists [§ 3.2, Remark]; we now want to construct it.

First we reduce the problem to the case $n = 1$, i.e. to $\mathbb{K}$ itself, by the following general lemma.

LEMMA. <u>Let the l.c.a. group</u> G <u>be the direct sum of</u> n <u>closed subgroups</u> G_j, $1 \le j \le n$, <u>i.e. such that each</u> $x\in G$ <u>can be represented uniquely in the form</u>

$$x = \sum_{j=1}^n x_j,\quad x_j\in G_j.$$

<u>Let a symmetric bicharacter</u> B <u>on</u> $G\times G$ <u>be given, i.e.</u> $B(x,y) = B(y,x)$ <u>for all</u> x, y <u>in</u> G. <u>Suppose that for</u> $j = 1,\dots,n$ <u>we have a</u> ψ_j <u>in</u> $Ch_2(G_j)$ <u>satisfying</u>

$$\psi_j(x_j+y_j) = \psi_j(x_j)\psi_j(y_j)B(x_j,y_j)\quad \text{for } x_j,\ y_j \text{ in } G_j.$$

<u>Then there is one and only one</u> $\psi\in Ch_2(G)$ <u>such that</u>

$$\psi(x+y) = \psi(x)\psi(y)B(x,y) \text{ for } x,\ y \text{ in } G, \text{ and } \psi|G_j = \psi_j,\ 1 \le j \le n.$$

<u>This</u> ψ <u>is given by</u>

$$\psi(x) = \prod_{j=1}^n \psi_j(x_j)\cdot \prod_{1\le j<k\le n} B(x_j,x_k) \text{ for } x = \sum_{j=1}^n x_j,\ x_j\in G_j.$$

The proof is a straightforward verification if $n = 2$. Then the lemma follows by induction.

We can apply the lemma to $\mathbb{K}^n$, with the canonical basis. Here the bicharacter is as in (7-1), with

$$\beta = (b_{jk})\in M_n(\mathbb{K}),\quad b_{kj}=b_{jk}.$$

Suppose that for arbitrary $c\in\mathbb{K}$ we can construct $\psi_c\in Ch_2(\mathbb{K})$ such that

$$\psi_c(u+v) = \psi_c(u)\psi_c(v)\chi(cuv),\ \text{for } u,\ v \text{ in } \mathbb{K}. \tag{7-3}$$

Then, by the lemma above, we can put

$$\psi_\beta(x) := \prod_{j=1}^n \psi_{b_{jj}}(x_j)\cdot\chi\Big(\sum_{1\le j<k\le n} b_{jk}x_jx_k\Big) \quad \text{for } x = (x_1,\dots,x_n)\in\mathbb{K}^n \tag{7-4}$$

to obtain a solution of (7-1). Note that (7-4) coincides with (7-2) in the case of char $\mathbb{K} \neq 2$, if we let $\psi_{b_{jj}}(u) := \chi(2^{-1}b_{jj}u^2)$, $u \in \mathbb{K}$, $1 \le j \le n$.

We now want to obtain, for $\mathbb{K} = \mathbb{F}_q((T))$, $q = 2^r$, a $\psi_c \in Ch_2(\mathbb{K})$ satisfying (7-3). This task falls naturally into two parts:

(i) To obtain a $\psi' \in Ch_2(\mathbb{K})$ such that

$$\psi'(x+y) = \psi'(x)\psi'(y)\chi(xy) \text{ for } x, y \text{ in } \mathbb{K}. \tag{7-5}$$

(ii) To obtain a $\psi'' \in Ch_2(\mathbb{K})$ such that

$$\psi''(x+y) = \psi''(x)\psi''(y)\chi(Txy) \text{ for } x, y \text{ in } \mathbb{K}. \tag{7-6}$$

Indeed, suppose that (i) and (ii) are solved. Given $c \in \mathbb{K}$, we have $c = c_1^2 + c_2^2T$ [§ 5.5, Lemma]. Then

$$\psi_c(x) := \psi'(c_1x)\psi''(c_2x), \quad x \in \mathbb{K},$$

is a solution to (7-3).

To solve problem (i), let

$$x = \sum_n a_nT^n, \quad y = \sum_n b_nT^n, \tag{7-7}$$

where for $n < 0$ at most finitely many a_n, b_n are $\neq 0$. Put

$$(x,y)' := \Big(\sum_{n\ge 0} a_nb_{-1-n}\Big)\cdot T^{-1}.$$

Then the term in xy containing T^{-1} is $(x,y)' + (y,x)'$. Further, $(x,y)'$ is additive in each variable x, y, and is a continuous function of the pair (x,y). It follows that ψ' defined by

$$\psi'(x) := \chi((x,x)'), \quad x \in \mathbb{K} \; [\chi = \chi_{\mathbb{K}}, \text{ cf. § 7.4, (IIb), with } p = 2],$$

is a solution of (7-5), and problem (i) is solved.

To solve problem (ii), let us put for x, y as in (7-7)

$$(x,y)'' := \Big(\sum_{n\ge 0} a_nb_{-2-n}\Big)\cdot T^{-1} = (x,Ty)'.$$

Then the term in Txy which contains T^{-1} is $(x,y)'' + (y,x)'' + a_{-1}b_{-1}T^{-1}$. Now let H_o be the subgroup of $\mathbb{K}$ given by

$$H_o := \{x_o \mid x_o = \sum_n a_nT^n \in \mathbb{K}, \text{ with } a_{-1} = 0\}.$$

Put

$$\psi_o''(x_o) := \chi((x_o,x_o)'') \text{ for } x_o \in H_o \; [\chi = \chi_{\mathbb{K}}].$$

Then we have

$$\psi_o''(x_o+y_o) = \psi_o''(x_o)\psi_o''(y_o)\chi(Txy) \text{ for } x_o, y_o \text{ in } H_o.$$

Next let $(e_j)_{1\le j\le r}$ be a basis of $\mathbb{F}_q$ over $\mathbb{F}_2$ ($q = 2^r$), so that every $a \in \mathbb{F}_q$ can be represented uniquely as $\sum_{j=1}^{r} n_je_j$ with $n_j = 0$ or 1. Let H_j, $1 \le j \le r$, be the additive group $\{0, e_jT^{-1}\}$ in $\mathbb{K} = \mathbb{F}_q((T))$. The additive group of $\mathbb{K}$ is the direct sum of the r+1 closed subgroups H_j, with $0 \le j \le r$. So we can apply the lemma above: it is enough to find ψ_j in $Ch_2(H_j)$, for $j = 0,1,\dots r$, such that

$$\psi_j''(x_j+y_j) = \psi_j''(x_j)\psi_j''(y_j)\chi(Tx_jy_j) \text{ for } x_j, y_j \text{ in } H_j. \qquad (7\text{-}8)$$

The case of H_o, i.e. $j = 0$, has already been settled. Now for H_j, $1 \le j \le r$, the condition (7-8) takes the form

$$1 = \psi_j''(x_j)^2\chi(Tx_j^{\,2}) , \quad x_j \in H_j.$$

Here χ assumes only the values ±1. Thus (7-8) will hold if we put, for $1 \le j \le r$, $\psi_j(0) := 1$ and $\psi_j(x_j) := 1$ or i according as $\chi(Tx_j^{\,2})$ is 1 or -1. Then the lemma above shows that, if we put

$$\psi''(x) := \prod_{j=0}^{r} \psi_j''(x_j)\cdot \prod_{0\le j<k\le r} \chi(Tx_jx_k) \text{ for } x = \sum_{j=0}^{r} x_j,\ x_j \in H_j,$$

we obtain a solution of (7-6). So problem (ii) is also solved.

Thus we have obtained a solution of the functional equation (7-1) in the case char $\mathbb{K} = 2$.

7.8 For a finite-dimensional vector space X over a local field $\mathbb{K}$ we define

$$B_o(X/\mathbb{K}) := \{(\sigma,\psi) \mid (\sigma,\psi) \in B_o(X),\ \sigma \in Sp(X/\mathbb{K})\}.$$

This is a proper subgroup of $B_o(X)$ unless $\mathbb{K}$ is $\mathbb{R}$ or $\mathbb{Q}_p$ [§ 7.5].

We come now to the main result in this §.

PROPOSITION. Let γ be in Is(X*, X/$\mathbb{K}$) and let $\mathbb{B}(X/\mathbb{K})$ be the group of unitary operators in $L^2(X)$ generated by the Weil operators W^γ, $M(\alpha)$, $\alpha \in Aut(X/\mathbb{K})$, $A(\psi)$, $\psi \in Ch_2(X/\mathbb{K})$, and the operators $c\cdot I$, $c \in \mathbb{T}$ [I:= identity operator in $L^2(X)$]. This group is clearly independent of γ. The group $\mathbb{B}(X/\mathbb{K})$ contains the Heisenberg group $\mathbb{A}(X)$ [§ 1.8] and

$$\pi_o(\mathbb{B}(X/\mathbb{K})) = B_o(X/\mathbb{K}),$$

where π_o is the mapping of $\mathbb{B}(X)$ into $B_o(X)$ defined in § 4 (1-7). Moreover, $\mathbb{B}(X/\mathbb{K})$ is the subgroup of all $S \in \mathbb{B}(X)$ such that $\pi_o(S)$ lies in $B_o(X/\mathbb{K})$.

COROLLARY. $\mathbb{B}(X/\mathbb{K}) = \mathbb{B}(X)$ if $\mathbb{K}$ is $\mathbb{R}$ or $\mathbb{Q}_p$ [cf.§ 7.3].

For the proof, we first have to show that $U(a,a^*)$ is in $\mathbb{B}(X/K)$ for any (a,a^*) in $X\times X^*$. Now $U(a,a^*) = U(0,a^*)U(a,0)$, and here $U(0,a^*) = A(\chi_{a^*})$ is in $\mathbb{B}(X/\mathbb{K})$ by definition. Also $U(a,0) = (W^\gamma)^{-1}A(\chi_{a\gamma^{-1}})W^\gamma$, as a slight calculation will show; hence $U(a,0) \in \mathbb{B}(X/\mathbb{K})$. Thus $\mathbb{B}(X/\mathbb{K})$ contains $\mathbb{A}(X)$. Next let $B_o'(X/\mathbb{K})$ be the subgroup of $B_o(X/\mathbb{K})$ generated by the following elements [cf. § 4 (1-8) for the notation]: w_o^γ, $m_o(\alpha)$, $\alpha \in \mathrm{Aut}(X/\mathbb{K})$, and $a_o(\psi_X)$, $\psi_X \in \mathrm{Ch}_2(X/\mathbb{K})$. Then, of course, $\pi_o(\mathbb{B}(X/\mathbb{K})) = B_o'(X/\mathbb{K})$. To show $B_o'(X/\mathbb{K}) = B(X/\mathbb{K})$, note that, as a consequence of Witt's theorem [§ 7.2], there is for every $(\sigma,\psi) \in B_o(X/\mathbb{K})$ an element $(\sigma,\psi') \in B_o'(X/\mathbb{K})$ <u>with the same</u> σ. Then $(\sigma,\psi) = (1,\chi')(\sigma,\psi')$, where χ' is a character of $X \times X^*$: $\chi' = \chi_{a^*} \otimes \chi_a$, $(a,a^*) \in X \times X^*$. In view of formula § 4 (1-10) (with -a in place of a), $(1,\chi')$ is also in $B_o'(X/\mathbb{K})$, whence the desired equality. Since $\mathbb{B}(X/\mathbb{K})$ contains the kernel of π_o [§ 4.6], the last assertion of the proposition follows.

The proposition above takes the place of Weil's Theorem 1 in the exposition of his theory given here [§ 8]; see also § 7.2, Remark.

REMARK. We note that the operators in $\mathbb{B}(X/\mathbb{K})$ leave $\mathcal{S}^1(X)$ invariant, more precisely, they induce automorphisms of $\mathcal{S}^1(X)$ as a Banach space. Here we obtain this result from the very definition of $\mathbb{B}(X/\mathbb{K})$ given in the proposition above, since it holds for the operators generating $\mathbb{B}(X/\mathbb{K})$ [§ 3.4]. For $\mathbb{K} = \mathbb{R}$ or $\mathbb{Q}_p$, this observation yields, in fact, the automorphism theorem in § 4.5 for G = X, by the corollary above.

<u>7.9</u> Let Γ be a closed subgroup of X. We can then consider $\mathbb{B}(X,\Gamma)$ and $B_o(X,\Gamma)$ [cf. the theorem in § 4.9]. We put

$$\mathbb{B}(X,\Gamma/\mathbb{K}) := \mathbb{B}(X,\Gamma) \cap \mathbb{B}(X/\mathbb{K}), \quad B_o(X,\Gamma/\mathbb{K}) := B_o(X,\Gamma) \cap B_o(X/\mathbb{K}).$$

Here we note a result, for later reference, which we obtain on combining the proposition in § 7.8 with the theorem in § 4.9 (cf. also Remark 1 in § 4.9): $\mathbb{B}(X,\Gamma/\mathbb{K})$ <u>can be obtained from</u> $B_o(X,\Gamma/\mathbb{K})$ <u>by a theta lifting</u>; in particular $\pi_o(\mathbb{B}(X,\Gamma/\mathbb{K})) = B_o(X,\Gamma/\mathbb{K})$.

REMARK. We also mention here: if we introduce the usual topology in $B_o(X/\mathbb{K})$, as a subgroup of $B_o(X)$ [§ 4.10], then $(\sigma,\psi) \longmapsto \sigma$ is a strict morphism of $B_o(X/\mathbb{K})$ onto $\mathrm{Sp}(X/\mathbb{K})$, and an isomorphism if char $\mathbb{K} \neq 2$ (cf. § 4.11, Remark).

§8. *The metaplectic group (local case); Segal continuity*

8.1 Consider an element σ of $Sp(X/\mathbb{K})$ [§ 7.3] and a character of the 2nd degree of the form $\chi\circ f$, $f\in Q(X\times X^*)$. Then $(\sigma,\chi\circ f)$ will belong to the group $B_0(X)$ if the following relation holds [cf. § 7 (1-2) and § 4.1]:

$$f(z_1+z_2) = f(z_1)+f(z_2)+b_1(z_1\sigma,z_2\sigma)-b_1(z_1,z_2),\quad z_j\in X\times X^*. \qquad (1\text{-}1)$$

It can be verified immediately that the elements $(\sigma,\chi\circ f)$ of $B_0(X)$ which satisfy (1-1) form a subgroup of $B_0(X)$. With this subgroup there is associated, in a natural way, the group of all pairs (σ,f), $\sigma\in Sp(X/\mathbb{K})$, $f\in Q(X\times X^*)$, for which the relation (1-1) holds, with the law of multiplication

$$(\sigma,f)\cdot(\sigma',f') := (\sigma\sigma',f + f'\circ\sigma)$$

which corresponds to the law of multiplication § 4 (1-6) in $B_0(X)$; it is easy to verify that we obtain a group. This is Weil's pseudosymplectic group, denoted by $Ps(X/\mathbb{K})$ or $Ps(X)$ for short. It is a locally compact group in the topology induced by the product topology of $Sp(X/\mathbb{K})\times Q(X\times X^*)$. We write $s = (\sigma,f)$ for the elements of $Ps(X)$.

REMARK. It is important for later purposes [§ 9] to observe that the definition of $Ps(X)$ applies, in fact, to a finite-dimensional vector space over any (commutative) field (Weil [25, n° 31, p. 180]).

Let for $\chi = \chi_{\mathbb{K}}$

$$s \longmapsto \mu(s) := (\sigma,\chi\circ f),\quad s = (\sigma,f)\in Ps(X),$$

be the natural morphism of $Ps(X)$ into $B_0(X)$. The kernel of μ consists of all $s = (\sigma,f)\in Ps(X)$ with $\sigma = 1$ and $\chi\circ f = 1$. But if $\sigma = 1$, then f is additive on $X\times X^*$, by (1-1), thus if char $\mathbb{K} \neq 2$ we have $f = 0$, i.e. μ is injective. In this case we also have: the relation (1-1) determines for $\sigma\in Sp(X/\mathbb{K})$ a unique $f\in Q(X\times X^*)$. It now follows that, when char $\mathbb{K} \neq 2$, $Ps(X)$ is isomorphic to $Sp(X/\mathbb{K})$ and may be identified with $\mu(Ps(X))$.

8.2 Put

$$\gamma = \gamma(s),\ \text{when}\ s = (\sigma,f)\in Ps(X),\ \sigma = \begin{pmatrix}\alpha & \beta\\ \gamma & \delta\end{pmatrix},$$

and let $\Omega(X)$ be the set of all $s\in Ps(X)$ for which $\gamma(s)$ is a $\mathbb{K}$-isomorphism of X^* onto X, i.e. det $\gamma(s) \neq 0$ (a condition independent of the choice of bases in X and X*). The set $\Omega(X)$ is fundamental in Weil's theory; it is the analogue of $\Omega_0(G)$ in § 4.3 and is an open set in Ps(X), since $s \longmapsto \det \gamma(s)$ is continuous on Ps(X). The main result about $\Omega(X)$ is analogous to the lemma in § 4.3. We introduce the following elements of Ps(X) [it is readily seen that they are elements of Ps(X)]:

(i) for $\alpha \in \mathrm{Aut}(X/\mathbb{K})$ we put $m(\alpha) := \left(\begin{pmatrix} \alpha & 0 \\ 0 & \alpha^{*-1} \end{pmatrix}, 0 \right)$;

(ii) for $f_X \in Q(X)$ we put $a(f_X) := \left(\begin{pmatrix} 1 & \beta \\ 0 & 1 \end{pmatrix}, f_X \otimes 1_{X^*} \right)$ if $f_X(x+y) = f_X(x)+f_X(y)+[x, y\beta]$ for x, y in X, where $\beta \in \mathrm{Mor}(X, X^*/\mathbb{K})$, $\beta^* = \beta$;

(iii) for $\gamma \in \mathrm{Is}(X^*, X/\mathbb{K})$ we put $w^{\gamma} := \left(\begin{pmatrix} 0 & -1 \\ 1 & 0 \end{pmatrix}, -b_0 \right)$, with b_0 as in § 7 (1-2). Note that for $\alpha' \in \mathrm{Aut}(X/\mathbb{K})$ we have $w^{\gamma\alpha'} = w^{\gamma} m(\alpha')$.

The elements of Ps(X) defined in (i)-(iii) are the analogues of the elements of $B_0(G)$ introduced in § 4 (1-8a,b,c).

PROPOSITION (Weil [25, p. 182]). Every $s = (\sigma, f) \in \Omega(X)$ has a unique representation

$$s = a(f_1) w^{\gamma} a(f_2), \quad \gamma \in \mathrm{Is}(X^*, X/\mathbb{K}), \quad f_j \in Q(X), \quad j = 1, 2, \tag{2-1}$$

and conversely. Here γ, f_1, f_2 are given by

$$\gamma = \gamma(s), \quad f_1(x) = f(x, -x\alpha\gamma^{-1}), \quad f_2(x) = f(0, x\gamma^{-1}).$$

The proof is the same as that of the proposition in § 4.3, with a mere change of notation.

COROLLARY. Let $\gamma \in \mathrm{Is}(X^*, X/\mathbb{K})$ be fixed. The mapping

$$(f_1, \alpha', f_2) \longmapsto a(f_1) w^{\gamma} m(\alpha') a(f_2), \quad (f_1, \alpha', f_2) \in Q(X) \times \mathrm{Aut}(X/\mathbb{K}) \times Q(X),$$

is a homeomorphism of $Q(X) \times \mathrm{Aut}(X/\mathbb{K}) \times Q(X)$ onto $\Omega(X)$.

Indeed, the mapping is continuous and it has an inverse, by the proposition above, which is also continuous: for, in (2-1) f_1, f_2 and $\alpha' := \gamma^{-1}\gamma(s)$ depend continuously on $s \in \Omega(X)$: this is most easily seen by means of a basis in X and the dual basis in X*.

8.3 The above corollary, combined with the neighbourhood lemma for Q(X) in § 5.7, yields at once the following neighbourhood lemma for Ps(X):

LEMMA. Let U be a neighbourhood of 1 in $\mathrm{Aut}(X/\mathbb{K})$. Let the quadratic form $f^{\circ} \in Q(X)$ be as in the neighbourhood lemma for Q(X) [§ 5.7]. Choose a fixed $\gamma \in \mathrm{Is}(X^*, X/\mathbb{K})$ and define the element s_1 of Ps(X) by

$s_1 := a(f^\circ)w^\gamma a(f^\circ)$.

Let $V_U(s_1;Ps)$ consist of all $s \in Ps(X)$ of the form

$s = a(f_1)w^\gamma m(\alpha')a(f_2)$, with f_1, f_2 in $V_U(f^\circ)$ [§ 5.7] and α' in U.

The set $V_U(s_1;Ps)$ is a neighbourhood of s_1 in Ps(X).

8.4 We now introduce Weil's metaplectic group Mp(X). We recall that $\mathbb{B}(X/\mathbb{K})$, as a group of unitary operators of $L^2(X)$ [§ 7.8], is a topological group in the strong operator topology. Consider the product $Ps(X)\times\mathbb{B}(X/\mathbb{K})$: we define Mp(X) as the subgroup of this product which consists of all pairs

$\underline{S} = (s,S) \in Ps(X)\times\mathbb{B}(X/\mathbb{K})$ such that $\mu(s) = \pi_0(s)$.

This is usually written

$Mp(X) := Ps(X)\times_{B_0(X)}\mathbb{B}(X/\mathbb{K})$

(fibre product of Ps(X) and $\mathbb{B}(X/\mathbb{K})$ over $B_0(X)$). We give Mp(X) the induced topology and we write

$s = \pi(\underline{S})$ if $\underline{S} = (s,S) \in Mp(X)$,

so π is a morphism (:= continuous homomorphism) of Mp(X) onto Ps(X), by § 7.8, Proposition, since $\mu(Ps(X))$ lies in $B_0(X/\mathbb{K})$.

REMARK. Since $\mu(Ps(X)) \subset B_0(X/\mathbb{K})$, the proposition in § 7.8 also shows that it amounts to the same if in the definition of Mp(X) we use $\mathbb{B}(X)$ instead of $\mathbb{B}(X/\mathbb{K})$. This is Weil's definition [25, n° 35, p. 185].

8.5 By the proposition in § 4.3 we have a lifting of $\Omega_0(X)$ into $\mathbb{B}(X)$; we denote this lifting by R_0. Then $R := R_0\circ\mu$ is a mapping of $\Omega(X)$ into $\mathbb{B}(X)$, more precisely into $\mathbb{B}(X/\mathbb{K})$: we have $\mu(\Omega(X)) \subset \Omega_0(X)$, and the definition of R_0 shows that $R_0(\mu(\Omega(X)) \subset \mathbb{B}(X/\mathbb{K})$. Explicitly we have, with the usual notation:

if $s = a(f_1)w^\gamma m(\alpha')a(f_2)$, then $R(s) = A(\chi\circ f_1)W^\gamma M(\alpha')A(\chi\circ f_2)$.

Note that R is a continuous mapping: the product depends continuously on each factor (which is a unitary operator) and the mappings $f \longmapsto A(\chi\circ f)$, $\alpha' \longmapsto M(\alpha')$ of Q(X) and $Aut(X/\mathbb{K})$ into $Aut(L^2(X))$ are continuous in the strong topology.

We can now lift $\Omega(X)$ to Mp(X) by

$$s \longmapsto (s,R(s)),\ s \in \Omega(X). \tag{5-1}$$

Thus $\pi((s,R(s)) = s$ and (5-1) is, of course, a homeomorphism of $\Omega(X)$ with its image [§ 1.12]. Moreover, we can apply § 1.13 to Mp(X) and the kernel of π which by § 4.6 is $\{(e,\tau\cdot I) \mid \tau \in T\}$ (e := neutral element of Ps(X), I := identity operator on $L^2(X)$). Hence we obtain that

$$(\tau, s) \longmapsto (s, \tau\cdot R(s)) = (e, \tau\cdot I)\cdot(s, R(s)) \qquad (5\text{-}2)$$

is a homeomorphism of $T\times\Omega(X)$ onto $\pi^{-1}(\Omega(X))$. Thus Mp(X) <u>is locally compact and</u> π <u>is</u> open, i.e. <u>a strict morphism of</u> Mp(X) <u>onto</u> Ps(X).

<u>8.6</u> By combining the results obtained so far, we can now establish the following basic property of the metaplectic group in the local case:

THEOREM. <u>The unitary representation</u> $\underline{S} = (s, S) \longmapsto S$ <u>of</u> Mp(X) <u>in</u> $L^2(X)$ <u>induces a continuous representation of</u> Mp(X) <u>in the Banach space</u> $\mathfrak{S}^1(X)$.

Explicitly:

Let $\underline{S} = (s, S) \in Mp(X)$, $\Phi \in \mathfrak{S}^1(X)$, and put, following Weil, $\underline{S}\Phi := S\Phi$. Then $\Phi \longmapsto \underline{S}\Phi$ is an automorphism of the Banach space $\mathfrak{S}^1(X)$, and the mapping $\underline{S} \longmapsto \underline{S}\Phi$ of Mp(X) into $\mathfrak{S}^1(X)$ is continuous. We call this the <u>Segal continuity of the metaplectic group</u>.

That $\Phi \longmapsto \underline{S}\Phi$ is an automorphism of the Banach space $\mathfrak{S}^1(X)$, stems from the corresponding fact for $\mathbb{B}(X/\mathbb{K})$, proved in § 4.5 for $\mathbb{B}(X)$.

To establish the continuity, it is enough to show, for fixed Φ in $\mathfrak{S}^1(X)$, that <u>for some</u> $\underline{S}_1 \in Mp(X)$ we have: given $\varepsilon > 0$, there is a nd. $V(\underline{S}_1; Mp)$ such that

$$\|\underline{S}\Phi - \Phi\|_{\mathfrak{S}} < \varepsilon \text{ for } \underline{S} \in V(\underline{S}_1; Mp).$$

To prove this, let U be any nd. of 1 in Aut(X/$\mathbb{K}$). Let $s_1 \in \Omega(X)$ and the nd. $V_U(s_1; Ps)$ be as in the neighbourhood lemma for Ps(X) [§ 8.3]. Let U' be a nd. of 1 in T (later U and U' will be chosen 'sufficiently small'). Then we put $\underline{S}_1 := (s_1, R(s_1))$ and

$$V_{U,U'}(\underline{S}_1; Mp) := \left\{ \underline{S} \mid \underline{S} = (s, S),\ s \in V_U(s_1; Ps),\ S = \tau\cdot R(s),\ \tau \in U' \right\}.$$

This is a nd. of $\underline{S}_1$ in Mp(X) (cf. (5-2)).

For $\underline{S} \in V_{U,U'}(\underline{S}_1; Mp)$ we can write

$$\underline{S}\Phi - \underline{S}_1\Phi = \tau\cdot[R(s)\Phi - R(s_1)\Phi] + (\tau - 1)\cdot R(s_1)\Phi.$$

We choose U' so small that

$$|\tau - 1|\cdot\|R(s_1)\Phi\|_{\mathfrak{S}} < \varepsilon/2 \quad \text{for all } \tau \in U'.$$

Then we want to prove that we can choose U so small that

$$\|R(s)\Phi - R(s_1)\Phi\|_{\mathfrak{S}} < \varepsilon/2 \quad \text{for all } s \in V_U(s_1; Ps). \qquad (6\text{-}1)$$

Let us now state the next step in the proof as a lemma. It shows the main role of the nd. $V_U(s_1;Ps)$ and the main idea of the proof, and it will be used again.

LEMMA. <u>Let</u> s <u>be in</u> $V_U(s_1;Ps)$; <u>put</u> $\psi_0 := \chi \circ f^\circ$ [cf. § 8.3, Lemma]. <u>Then</u> [cf. § 8.5]

$$R(s) = A(\psi_0 \circ \alpha_1) W^{\gamma} M(\alpha') A(\psi_0 \circ \alpha_2) \text{ with } \alpha', \alpha_1, \alpha_2 \text{ in U.} \qquad (6\text{-}2)$$

First let char $\mathbb{K} \neq 2$. Then $s \in V_U(s_1;Ps)$ has by definition [§ 8.3] the form

$$s = a(f^\circ \circ \alpha_1) w^{\gamma} m(\alpha') a(f^\circ \circ \alpha_2) \text{ with } \alpha', \alpha_1, \alpha_2 \text{ in U.}$$

Thus

$$\mu(s) = a_0(f^\circ \circ \alpha_1) w_0^{\gamma} m_0(\alpha') a_0(f^\circ \circ \alpha_2) \qquad (6\text{-}3)$$

and $R(s) = R_0(\mu(s))$ is as stated, by the very definition of R(s) in § 8.5.

Now let char $\mathbb{K} = 2$. Then $s \in V_U(s_1;Ps)$ is, by definition [§ 8.3], of the form

$$s = a(f^\circ \circ \alpha_1 + f_1') w^{\gamma} m(\alpha') a(f^\circ \circ \alpha_2 + f_2')$$

with

$$\alpha', \alpha_1, \alpha_2 \text{ in U}, f_1', f_2' \text{ in } Q'(X).$$

But for such an s we have: $\mu(s)$ is also given by (6-3). For, μ is a morphism of Ps(X) into $B_0(X)$ [§ 8.1] and

$$a(f^\circ \circ \alpha_j + f_j') = a(f^\circ \circ \alpha_j) a(f_j'), \; j = 1, 2.$$

As f_j' is in Q'(X), it is additive; also $f_j'(x)$ is a square in $\mathbb{K}$ ($x \in \mathbb{K}$), so that $\chi(f_j'(x)) = 1$ [§ 7.4 (IIb), § 5.5]. Hence $a(f_j')$ lies in the kernel of μ! So $\mu(s)$ is still given by (6-3) and (6-2) still holds.

By the lemma above, to show (6-1) amounts to showing that we can choose a nd. U of 1 in Aut(X/$\mathbb{K}$) so small that

$$\| A(\psi_0 \circ \alpha_1) W^{\gamma} M(\alpha') A(\psi_0 \circ \alpha_2) \Phi - A(\psi_0) W^{\gamma} A(\psi_o) \Phi \|_{\mathfrak{S}} < \varepsilon/2$$

for all $\alpha', \alpha_1, \alpha_2$ in U.

Such a choice of U is indeed possible: this is precisely the content of the continuity lemma for Weil operators [§ 3.8, Lemma 1].

Thus the proof of the theorem is complete.

<u>8.7</u> The metaplectic group Mp(X) has another important property with regard to $\mathfrak{S}^1(X)$ which we note here for later reference.

PROPOSITION. Let K be a compact set in Mp(X), in particular a compact neighbourhood of the neutral element. There is a constant C(K) such that for all $\Phi \in \mathfrak{S}^1(X)$

$$\|\underline{S}\Phi\|_{\mathfrak{S}} \le C(K)\cdot\|\Phi\|_{\mathfrak{S}} \quad \text{if } \underline{S} = (s,S) \text{ lies in } K.$$

To prove this, we combine the lemma in § 8.6 with Lemma 2 in § 3.8 for G = X. Then we obtain a nd. of 1 in Mp(X) for which the assertion holds, and by finitely many translations the proposition follows.

The proposition above, combined with the theorem in § 8.6, shows: $\underline{S}\Phi$ is a continuous function of $(\underline{S},\Phi) \in Mp(X)\times\mathfrak{S}^1(X)$.

8.8 We can say more in the case char $\mathbb{K} \neq 2$. Then we can put, in analogy to § 4.11, with b_1 as in § 7 (1-2),

$$f_\sigma(z) := (1/2)b_1(z\sigma,z\sigma) - (1/2)b_1(z,z), \quad z \in X\times X^*, \tag{8-1}$$

and

$$\psi_\sigma := \chi\circ f_\sigma, \quad \sigma \in Sp(X/\mathbb{K}).$$

The elements (σ,ψ_σ), $\sigma \in Sp(X/\mathbb{K})$, form a subgroup of $B_0(X/\mathbb{K})$ which is isomorphic, as a topological group [cf. § 4.10], to $Sp(X/\mathbb{K})$, under $(\sigma,\psi_\sigma) \longmapsto \sigma$: for, $\sigma \longmapsto \psi_\sigma$ is continuous, so that § 1.12 applies. This subgroup is simply $\mu(Ps(X))$: indeed, Ps(X) consists here of all elements (σ,f_σ), $\sigma \in Sp(X/\mathbb{K})$, with f_σ as in (8-1), and Ps(X) is, of course, provided with the product topology induced by $Sp(X/\mathbb{K})\times Q(X\times X^*)$ [cf. in this context § 5.6]. Thus Ps(X) is also isomorphic to $Sp(X/\mathbb{K})$ under $(\sigma,f_\sigma) \longmapsto \sigma$, since $\sigma \longmapsto f_\sigma$ is continuous, so that § 1.12 applies again. Hence Ps(X) and $\mu(Ps(X))$ are isomorphic as topological groups under the mapping $(\sigma,f_\sigma) \longmapsto (\sigma,\chi\circ f_\sigma)$. We can now show:

If char $\mathbb{K} \neq 2$, then Mp(X) is isomorphic, as a topological group, to the subgroup $\pi_0^{-1}(\mu(Ps(X)))$ of $\mathbb{B}(X/\mathbb{K})$.

This results from the following general argument. Let G_0, G_1, G_2 be topological groups. Let π_1 be an isomorphism (of topological groups) of G_1 into G_0, and π_2 a morphism of G_2 into G_0 (not necessarily strict). Put $G_1' := \pi_1(G_1)$. The fibre product $G_1\times_0 G_2$ of G_1 and G_2 over G_0 is isomorphic to $\pi_0^{-1}(G_1')$. This is seen as follows. The isomorphism π_1 induces an isomorphism $(x,y) \longmapsto (\pi_1(x),y)$ of $G_1\times G_2$ onto $G_1'\times G_2$. This isomorphism maps the subgroup

$$G_1\times_0 G_2 := \{(x,y) \mid (x,y) \in G_1\times G_2,\ \pi_1(x) = \pi_2(y)\}$$

onto the subgroup

$$\{(x',y) \mid x' \in G_1',\ y \in G_2,\ \pi_2(y) = x'\}$$

of $G_0 \times G_2$. Putting $H := \pi_2^{-1}(G_1')$, we can write this as the subgroup

$$\{(x',y) \mid x' \in G_1',\ y \in H,\ \pi_2(y) = x'\}$$

of $G_0 \times H$. But this last subgroup is isomorphic to H [§ 1.12].

To prove the above assertion on the metaplectic group, we can apply this argument to $G_0 = B_0(X)$, $G_1 := Ps(X)$, $G_2 := \mathbb{B}(X/\mathbb{K})$, $\pi_1 := \mu$, $\pi_2 :=$ = restriction of π_0 to $\mathbb{B}(X/\mathbb{K})$. Recall, in this context, the proposition in § 7.8. We note that the continuity of π_0 is a rather general fact [§ 1.6, Prop. 2].

8.9 We further have: <u>if</u> char $\mathbb{K} \neq 2$, <u>then</u> $\mathbb{B}(X/\mathbb{K})$ <u>is locally compact and the restriction of</u> π_0 <u>to</u> $\mathbb{B}(X/\mathbb{K})$ <u>is a strict morphism of</u> $\mathbb{B}(X/\mathbb{K})$ <u>onto</u> $B_0(X/\mathbb{K})$. <u>Also,</u> $\mathbb{B}(X/\mathbb{K})$ <u>has the Segal continuity property</u>.

This is proved in three parts.

1. We show exactly as in § 4.11, writing $A_0(X)$ in place of $A_0(G)$: $B_0(X/\mathbb{K}) = A_0(X) \cdot \mu(Ps(X))$, $A_0(X) \cap \mu(Ps(X)) = (1,1)$ [=neutral element]. Here the normal subgroup $A_0(X)$ is isomorphic to $X \times X^*$ (as a topological group). The topology of $B_0(X/\mathbb{K})$ is the product topology of these subgroups; in particular, $B_0(X/\mathbb{K})$ is locally compact.

2. The set $\Omega_* := A_0(X) \cdot \Omega_0(X)$, with $\Omega_0(X) = \mu(\Omega(X))$, is open in $B_0(X/\mathbb{K})$ and we can define a lifting R_* of it to $\mathbb{B}(X/\mathbb{K})$ [cf. § 4 (11-5), (1-10)]:

$$R_*((1, \chi_{a^*} \otimes \bar{\chi}_a) \cdot (\sigma, \chi \circ f_\sigma)) := U(a,a^*) \cdot R_0(\sigma, \chi \circ f_\sigma),$$
$$(a,a^*) \in X \times X^*,\quad (\sigma, \chi \circ f_\sigma) \in \Omega_0(X).$$

This lifting is clearly continuous. It follows [§ 1.13] that <u>the open set</u> $\pi_0^{-1}(\Omega_*)$ <u>in</u> $\mathbb{B}(X/\mathbb{K})$ <u>is homeomorphic to</u> $T \times (X \times X^*) \times \Omega_0(X)$ (where we may replace $\Omega_0(X)$ by $\Omega(X)$, of course). Thus $\mathbb{B}(X/\mathbb{K})$ is indeed locally compact, and the restriction of π_0 to $\mathbb{B}(X/\mathbb{K})$ is open.

3. From step 2 we now obtain: $\mathbb{B}(X/\mathbb{K})$ has the Segal property. This follows very simply from the Segal property of $Mp(X) \cong \pi_0^{-1}(\mu(Ps(X)))$, since the operators $U(a,a^*)$ are isometric on $\mathfrak{S}^1(X)$ and the mapping $(a,a^*) \longmapsto U(a,a^*)\Phi$ of $X \times X^*$ into $\mathfrak{S}^1(X)$ is continuous.

What happens when char $\mathbb{K} = 2$ is an open question.

In this context, another problem should be mentioned. Igusa [9, Lemma 4, p. 195] has shown for an elementary group E that $\mathbb{B}(E)$ is a Lie group. The question naturally arises whether in the case $E^* = E$ the Segal continuity property also holds for $\mathbb{B}(E)$. But this, it seems, can only be decided by methods different from those used here.

§9. *The metaplectic group and Segal continuity in the adelic case*

<u>9.1</u> It will be useful to recall briefly the main facts concerning adeles; for details cf. Weil [26, Ch. IV], Cassels-Fröhlich [6, Chs. II, XV], Serre [23], Bourbaki [1]. The reader should observe that there are some slight differences in notation and terminology between Weil's memoir [25] and his book [26], and likewise between the various authors.

We consider two kinds of fields:

(i) algebraic number fields, or <u>number fields</u> for short, i.e. the field $\mathbb{Q}$ and its finite algebraic extensions;

(ii) algebraic function fields, or simply <u>function fields</u>, i.e. the fields (which are algebraically isomorphic to) $\mathbb{F}_p(T)$, the field of rational functions in one indeterminate T, with coefficients in the field $\mathbb{F}_p := \mathbb{Z}/p\mathbb{Z}$, where p is a prime, or finite algebraic extensions of $\mathbb{F}_p(T)$.

These may be called <u>arithmetical fields</u>; they are usually called global fields, but this terminology is avoided by Weil. They will be denoted by the gothic letter $\mathfrak{k}$.

Let ι be an embedding (:= isomorphism) of $\mathfrak{k}$ into a local field $\mathbb{K}$ such that $\iota(\mathfrak{k})$ is dense in $\mathbb{K}$. We call $(\iota,\mathbb{K})$ a completion of $\mathfrak{k}$. Two completions $(\iota,\mathbb{K})$, $(\iota',\mathbb{K}')$ are called equivalent if there is an isomorphism ρ of $\mathbb{K}$ onto $\mathbb{K}'$ such that $\iota' = \rho\circ\iota$. A <u>place</u> of $\mathfrak{k}$ is defined as an equivalence class of completions. A place is said to be <u>infinite</u> if $\mathbb{K}$ is $\mathbb{R}$ or $\mathbb{C}$ (and is then called <u>real</u> or <u>complex</u>, respectively), <u>finite</u> if $\mathbb{K} \neq \mathbb{R}$ or $\mathbb{C}$. We denote a place by v, for the following reason: the function

$$x \longmapsto |x|_v := \begin{cases} \operatorname{mod}_{\mathbb{K}}(\iota(x)) \text{ if } \mathbb{K} \neq \mathbb{C}, \\ \{\operatorname{mod}_{\mathbb{K}}(\iota(x))\}^{1/2} \text{ if } \mathbb{K} = \mathbb{C}, \end{cases}$$

is a 'normalized' absolute value $|\ |_v$ on $\mathfrak{k}$ and only depends on the place. Thus $|\ |_v$ characterizes the place; $\mathbb{K}$ may be considered as the completion of $\mathfrak{k}$ with respect to the metric $|x-y|_v$, and denoted by $\mathfrak{k}_v$. To a finite place there corresponds an ultrametric absolute value, and conversely.

An arithmetical field has countably many places; a number field has finitely many infinite places, a function field has only finite places. For a finite place v, we denote by $\mathfrak{o}_v$ the ring of integers of $\mathfrak{k}_v$, consisting of the elements $x \in \mathfrak{k}_v$ such that $|x|_v \le 1$. Thus $\mathfrak{o}_v$ is a compact, open subgroup of the additive group of $\mathfrak{k}_v$.

Given an arithmetical field $\mathfrak{k}$, we define the <u>adele group</u> $\mathfrak{k}_A$ as the additive group of all 'sequences' $(x_v)_v$, with $x_v \in \mathfrak{k}_v$, v ranging over all places of $\mathfrak{k}$, and $x_v \in \mathfrak{o}_v$ for <u>almost all</u> v, i.e. for all except possibly finitely many v. This is a locally compact group if we take

$$\prod_{v\in P_\infty} \mathfrak{k}_v \times \prod_{v\notin P_\infty} \mathfrak{o}_v \ , \quad P_\infty := \text{set of finite places of } \mathfrak{k},$$

with the product topology, as an open subgroup of $\mathfrak{k}_A$. We can represent $\mathfrak{k}_A$ as the union of open subgroups $\mathfrak{k}^\circ_P$, thus as an inductive limit:

$$\mathfrak{k}_A = \operatorname*{ind.lim}_{P\supset P_\infty} \mathfrak{k}^\circ_P \ , \quad \mathfrak{k}^\circ_P := \prod_{v\in P} \mathfrak{k}_v \times \prod_{v\notin P} \mathfrak{o}_v \ ,$$

where $\mathfrak{k}^\circ_P$ has the product topology and P ranges over all finite sets of places of $\mathfrak{k}$ containing P_∞. We may replace here '(additive) group' by 'ring', multiplication of 'sequences' being defined term by term.

Now let X be a finite-dimensional vector space over $\mathfrak{k}$. By choosing a basis of X over $\mathfrak{k}$, say X°, we can represent X as $\mathfrak{k}^n$, and thus embed it in $\mathfrak{k}_v^n$, for each place v. We then write X_v instead of $\mathfrak{k}_v^n$, since the choice of the basis is irrelevant (one can also give the 'abstract' definition $X_v := X \otimes_{\mathfrak{k}} \mathfrak{k}_v$). We can then introduce the <u>adele group</u> X_A as $\mathfrak{k}_A^n$; the notation indicates that X_A is independent of the choice of the basis X° of X (in abstract terms, $X_A := X \otimes_{\mathfrak{k}} \mathfrak{k}_A$). For a finite place v we denote by X°_v the compact, open subgroup of X_v generated by X° over $\mathfrak{o}_v$. We may represent X_A as an inductive limit of open subgroups:

$$X_A = \operatorname*{ind.lim}_{P\supset P_\infty} X^\circ_P, \quad X^\circ_P := \prod_{v\in P} X_v \times \prod_{v\notin P} X^\circ_v \ ,$$

where X°_P has the product topology and P ranges over all finite sets of places containing P_∞.

REMARK. If we choose another basis X° of X over $\mathfrak{k}$, then X°_v remains unchanged for almost all v.

We can embed (the additive group of) $\mathfrak{k}$ in $\mathfrak{k}_A$ by

$$\xi \longmapsto (\xi_v)_v \ , \ \xi_v := \xi \text{ for all places v of } \mathfrak{k}. \tag{1-1}$$

Then $\mathfrak{k}$ is a <u>discrete</u> subgroup of $\mathfrak{k}_A$ and $\mathfrak{k}_A/\mathfrak{k}$ is <u>compact</u>. Likewise we obtain a natural embedding of X into X_A: by choosing a basis X° of X, this amounts to the natural embedding of $\mathfrak{k}^n$ in $\mathfrak{k}_A^n$; as it is clearly independent of the choice of X°, we denote the image of X under this embedding by X_k. Again, X_k is a <u>discrete</u> subgroup of X_A and X_A/X_k is compact.

We also note that X_A is a vector space over k, multiplication by elements of k being defined in the obvious way. X_k is then isomorphic to X as a vector space.

<u>9.2</u> Any character $\chi = \chi_A$ of k_A is of the form

$$\chi = \otimes_v \chi_v \text{ , i.e. } \chi(x) = \prod_v \chi_v(x_v) \quad \text{for } x = (x_v)_v \in k_A \text{ ,} \tag{2-1}$$

where v ranges over all places of k and χ_v is a character of (the additive group of) k_v which is trivial (i.e. = 1) on o_v <u>for almost all</u> v so that the product is, in fact, a finite one for each x. We can choose χ so that it is trivial on k (embedded in k_A as in (1-1)), but not trivial on k_A ; e.g. this will be so if we choose each χ_v as in § 7.4. For any such χ we have:

$$\text{for almost all v, say for } v \notin P_o \ (P_o \supset P_\infty), \quad \chi_v(x_v) = 1 \Leftrightarrow x_v \in o_v. \tag{2-2}$$

Let again X be an n-dimensional vector space over k, X^* its algebraic dual, $[x,x^*]$ the value of $x^* \in X^*$ at $x \in X$. We can then consider $X^*_A := (X^*)_A$. X^*_A will also be represented by k_A^n if we introduce a basis in X^*, and is a vector space over k. We can embed X^* as a discrete subgroup X^*_k in X^*_A, with compact quotient; X^*_k is isomorphic to X^* as a vector space, X^*_A being a vector space over k.

<u>We may identify</u> (the additive group of) X^*_A <u>with the dual group, in the sense of harmonic analysis, of</u> (the additive group of) X_A. <u>The duality can be realized as follows</u>. We choose any non-trivial character χ of k_A trivial on k and put

$$\langle x,x^* \rangle := \chi([x,x^*]), \quad x \in X_A, \quad x^* \in X^*_A \text{ ,}$$

where we use again $[x,x^*]$ to denote the natural extension of the original bilinear form above, defined on $X \times X^*$, to a k-bilinear mapping of $X_A \times X^*_A$ into k_A. We also have: <u>the subgroup of</u> (the additive group of) X^*_A <u>associated</u> ('<u>orthogonal</u>') <u>to</u> $X_k \subset X_A$ <u>by duality is</u> (the additive group of) X^*_k. It is usual to omit the distinction between vector spaces and their additive groups, the meaning being clear from the context.

Given a basis X° of X over k, let, for simplicity, $(X^*)^\circ$ be the basis of X^* dual to X°. For finite v let $(X^*)^\circ_v$ be the compact, open subgroup of X^*_v generated by $(X^*)^\circ$ over o_v; thus

$$x^* \in (X^*)^\circ_v \Leftrightarrow [x,x^*] \in o_v \text{ for all } x \in o_v \text{ .}$$

Then (2-2) shows that

for $v \notin P_o$ we have $(X^*)^\circ_v = (X^\circ_v)_*$ (orthogonal subgroup). (2-3)

This fact will be of importance later on.

9.3 Consider now the subgroup $Sp(X/\mathfrak{k})$ of $Aut(X\times X^*/\mathfrak{k})$ [cf. § 7.1]. Its 'adelization' $Sp(X)_A$ may be described as follows, if we introduce a basis X° of X over $\mathfrak{k}$ and the dual basis $(X^*)^\circ$ of X^*. For each place v of $\mathfrak{k}$ we write $Sp(X)_v$ for $Sp(X_v/\mathfrak{k}_v)$. For finite v, we let $Sp(X)^\circ_v$ denote the subgroup of all σ_v in $Sp(X)_v$ which induce an automorphism of $X^\circ_v\times(X^*)^\circ_v$; relatively to the bases X°, $(X^*)^\circ$, such a σ_v is (represented by) a matrix in $Sp_{2n}(\mathfrak{o}_v)$ ($n := \dim(X/\mathfrak{k})$), i.e. by a matrix in $Sp_{2n}(\mathfrak{k}_v)$ having entries in $\mathfrak{o}_v$. Hence $Sp(X)^\circ_v$ is a compact, open subgroup of $Sp(X)_v$. Denote again by P any finite set of places containing the set P_∞ of infinite places, and put

$$Sp(X)^\circ_P := \prod_{v\in P} Sp(X)_v \times \prod_{v\notin P} Sp(X)^\circ_v ,$$

which is a locally compact group with the usual product topology. Also, if $P' \supset P$, then $Sp(X)^\circ_P$ is open in $Sp(X)^\circ_{P'}$. We can represent $Sp(X)_A$ as an inductive limit:

$$Sp(X)_A := \operatorname{ind.lim}_{P\supset P_\infty} Sp(X)^\circ_P .$$

The elements of $Sp(X)_A$ are thus 'sequences' $(\sigma_v)_v$, $\sigma_v\in Sp(X)_v$ for all v, $\sigma_v \in Sp(X)^\circ_v$ for almost all v, with multiplication term by term and the inductive limit topology.

The group $Sp(X)_A$, as introduced above, does not depend on the basis X°. A change of basis in $\mathfrak{k}^n$ determined by $\rho\in GL_n(\mathfrak{k})$ gives [cf. § 7 (2-2)], if we put $\sigma_m(\rho) := \begin{pmatrix}\rho & 0\\ 0 & {}^t\rho^{-1}\end{pmatrix} \in Sp_{2n}(\mathfrak{k})$:

$$\begin{pmatrix}\alpha' & \beta'\\ \gamma' & \delta'\end{pmatrix} = \sigma_m(\rho)^{-1}\begin{pmatrix}\alpha & \beta\\ \gamma & \delta\end{pmatrix}\sigma_m(\rho) , \text{ for } \begin{pmatrix}\alpha & \beta\\ \gamma & \delta\end{pmatrix} \in Sp_{2n}(\mathfrak{k}_v).$$

This is an inner automorphism of $Sp_{2n}(\mathfrak{k}_v)$ and, for almost all v, it leaves $Sp_{2n}(\mathfrak{o}_v)$ invariant. Thus $Sp(X)_A$ is indeed independent of the basis X° of X which justifies the notation. For the canonical basis of $\mathfrak{k}^n$ we may of course write $Sp_{2n}(\mathfrak{k})_A$.

Next we consider the pseudosymplectic group $Ps(X) = Ps(X/\mathfrak{k})$ [cf. § 8.1, Remark]. Its adelization $Ps(X)_A$ may be described as follows. We write $Ps(X)_v$ for $Ps(X_v/\mathfrak{k}_v)$ and, for finite v, we let $Ps(X)^\circ_v$ be the compact and open subgroup of all (σ_v, f_v) such that (i) σ_v induces an automorphism of $X^\circ_v\times(X^*)^\circ_v$, (ii) f_v assumes on $X^\circ_v\times(X^*)^\circ_v$ values in $\mathfrak{o}_v$. This means that, relatively to the bases X°, $(X^*)^\circ$, σ_v has the same matrix representation as above, i.e. by an element of $Sp_{2n}(\mathfrak{o}_v)$, and f_v is represented by a quadratic form on $\mathfrak{k}_v^{2n}$ with coefficients in $\mathfrak{o}_v$. Then, similarly as above, we have a representation of $Ps(X)_A$:

$$Ps(X)_A = \operatorname{ind.lim}_{P \supset P_\infty} Ps(X)_P^\circ \text{ , where } Ps(X)_P^\circ := \prod_{v \in P} Ps(X)_v \times \prod_{v \notin P} Ps(X)_v^\circ .$$

The elements of the locally compact group $Ps(X)_A$ are the 'sequences' $s = (s_v)_v = ((\sigma_v, f_v))_v$ with s_v in $Ps(X)_v$ for all v, and in $Ps(X)_v^\circ$ for almost all v, and with term by term multiplication.

We show that $Ps(X)_A$ is independent of the choice of the basis X°. A change of basis as above induces, for each v, an inner automorphism of $Ps_{2n}(\mathfrak{k}_v)$ given by

$$(\sigma, f) \longmapsto (\sigma_m(\rho)^{-1}\sigma\sigma_m(\rho), f\circ\sigma_m(\rho)^{-1}) = (\sigma_m(\rho),0)^{-1}(\sigma,f)(\sigma_m(\rho),0),$$

and the stated independence then follows as before, whence the notation. For the canonical basis in $\mathfrak{k}^n$ we may write $Ps_{2n}(\mathfrak{k})_A$.

The group $Ps(X)_A$ plays the same role for X_A as the 'local' pseudo-symplectic group does in the local case [§ 8], as we shall see.

Finally, we remark that, when char $\mathfrak{k} \neq 2$, then $((\sigma_v, f_v))_v \longmapsto (\sigma_v)_v$ is an isomorphism of $Ps(X)_A$ onto $Sp(X)_A$. We shall return to this case in § 9.16.

<u>9.4</u> Before we consider $\mathcal{G}^1(X_A)$, we discuss $\mathcal{G}^1(X_P)$, where $P \supset P_\infty$ is as usual and

$$X_P := \prod_{v \in P} X_v .$$

Now $\mathcal{G}^1(X_P)$ has a tensor representation in terms of $\mathcal{G}^1(X_v)$, $v \in P$ [§ 2.6 (ii)]: the functions Φ_P in $\mathcal{G}^1(X_P)$ are precisely those of the form

$$\Phi_P = \sum_{n\geq 1} \Phi_{Pn}^\otimes , \quad \Phi_{Pn}^\otimes := \bigotimes_{v\in P} \Phi_{vn} , \text{ i.e. } \Phi_{Pn}^\otimes((x_v)_{v\in P}) = \prod_{v\in P} \Phi_{vn}(x_v), \tag{4-1a}$$

where

$$\Phi_{vn} \in \mathcal{G}^1(X_v),\ n \geq 1,\ \sum_{n\geq 1}\prod_{v\in P} \|\Phi_{vn}\|_{(v)} < \infty , \tag{4-1b}$$

$\| \ \|_{(v)}$ denoting a norm in $\mathcal{G}^1(X_v)$, chosen once for all. The norm in $\mathcal{G}^1(X_P)$ can then be defined by

$$\|\Phi_P\|_{(P)} := \inf \sum_{n\geq 1}\prod_{v\in P} \|\Phi_{vn}\|_{(v)} ,$$

the infimum being taken over all tensor representations (4-1a,b) of Φ. We note that the series in (4-1a) also converges uniformly on X_P, and in $L^2(X_P)$, since on each X_v the Segal norm dominates both the uniform norm and the L^2-norm.

The finite sums of functions of the type

$$\Phi^{\otimes} := \underset{v\in P}{\otimes}\Phi_v \ , \ \Phi_v \in \mathfrak{S}^1(X_v),$$

form a <u>dense linear subspace</u> $\mathfrak{S}(X_P)$ of $\mathfrak{S}^1(X_P)$.

Given operators $S_v \in \mathbb{B}(X_v/f_v)$, $v \in P$, there is a unique automorphism S_P of $L^2(X_P)$, their <u>tensor product</u>, denoted by

$$S_P = \underset{v\in P}{\otimes} S_v \ , \tag{4-2}$$

such that

$$S_P\Phi^{\otimes} = \underset{v\in P}{\otimes} S_v\Phi_v \ \text{for } \Phi^{\otimes} \text{ as above.} \tag{4-3}$$

Indeed, as is readily seen, S_P is uniquely defined by (4-3) [note that the representation of $\Phi^{\otimes}$ involves a certain non-uniqueness!], and can then be uniquely extended, by additivity, to an automorphism of $\mathfrak{S}(X_P)$, viewed as a subspace of the Hilbert space $L^2(X_P)$. Now $\mathfrak{S}(X_P)$ is dense in $L^2(X_P)$, being dense in $\mathfrak{S}^1(X_P)$ which is dense in $L^2(X_P)$. Thus S_P can be extended uniquely, by continuity, to an automorphism of the space $L^2(X_P)$.

<u>If</u> $\Phi_P \in \mathfrak{S}^1(X_P)$, <u>then</u> $S_P\Phi$ <u>is also in</u> $\mathfrak{S}^1(X_P)$. <u>If</u> Φ <u>has the tensor representation</u> (4-1), <u>then</u>

$$S_P\Phi_P = \sum_{n\geq 1} S_P\Phi^{\otimes}_{Pn} \tag{4-4}$$

<u>is a tensor representation of</u> $S_P\Phi_P$, <u>and</u>

$$|S_P| \leq \prod_{v\in P} |S_v| \quad [\text{cf. } (4\text{-}2)], \tag{4-5}$$

where $|\ |$ denotes the Segal operator norm for each of the operators S_v, S_P. This fact, and its proof, are an obvious extension of the case treated in § 3.4.

We can now discuss the Segal algebra $\mathfrak{S}^1(X_A)$. We suppose that the local characters have been chosen as in § 7.4 (cf. § 9.2). We choose the local Haar measures and the local Segal norms $\|\ \|_{(v)}$ as follows. For each infinite place v we take any Haar measure on X_v and the dual Haar measure on X^*_v; then we choose any compact nd. V^*_v of 0 in X^*_v such that V^*_v has dual Haar measure 1. This determines the Segal norm for all infinite places. For any finite place v we choose a basis $X°$ of X over f and take that Haar measure on X_v for which the compact, open subgroup $X°_v$ has measure 1; on X^*_v we take again the dual Haar measure. We put $V^*_v := (X°_v)_*$ so that V^*_v has dual Haar measure 1 (cf. e.g. [16, Ch.5, § 5.4, Example, p.122]). This determines the Segal norm for all finite places v. We recall that $(X°_v)_*$ <u>coincides with</u> $(X^*)°_v$ <u>for almost</u>

<u>all</u> v, <u>more precisely for</u> $v \notin P_o$ [cf. (2-2), (2-3)].

With this choice of Haar measures and of V_v^* we then have

$$\|\Phi_v\|_\infty \le \|\Phi_v\|_{(v)} \quad \text{for all } \Phi_v \in \mathfrak{S}^1(X_v) \text{ [cf. § 2 (3-5)].} \tag{4-6}$$

Also, for all finite places v, the characteristic function φ_v° of X_v° is in $\mathfrak{S}^1(X_v)$, since $(\varphi_v^\circ)^\wedge$ is the characteristic function of $(X_v^\circ)_*$, and it also follows that $\|\varphi_v^\circ\|_{(v)} = 1$ [cf. § 2.3. Remark 1].

Now let

$$\Phi_v \in \mathfrak{S}^1(X_v) \text{ for all } v, \quad \Phi_v = \varphi_v^\circ \text{ for almost all } v, \tag{4-7a}$$

and put

$$\Phi^\otimes := \bigotimes_v \Phi_v \text{ , i.e. } \Phi^\otimes((x_v)_v) := \prod_v \Phi_v(x_v) \text{ .} \tag{4-7b}$$

$\mathfrak{S}^1(X_A)$ <u>consists of all functions</u> Φ <u>on</u> X_A <u>which have a tensor representation</u>

$$\Phi = \sum_{n\ge1} \Phi_n^\otimes \text{ , } \quad \Phi_n^\otimes = \bigotimes_v \Phi_{vn} \text{ ,} \tag{4-8a}$$

<u>where for each</u> n

$$\Phi_{vn} \in \mathfrak{S}^1(X_v) \text{ for all } v, \; \Phi_{vn} = \varphi_v^\circ \text{ for almost all } v \tag{4-8b}$$

(note that here the 'almost all' may vary with n !) <u>and</u>

$$\sum_{n\ge1} \prod_v \|\Phi_{vn}\|_{(v)} < \infty \text{ .} \tag{4-8c}$$

<u>The norm in</u> $\mathfrak{S}^1(X_A)$ <u>can be defined by</u>

$$\|\Phi\|_{(A)} := \inf \sum_{n\ge1} \prod_v \|\Phi_{vn}\|_{(v)} \text{ ,}$$

<u>the infimum being taken over all tensor representations</u> (4-8) <u>of</u> Φ.

The proof of this is a straightforward extension of the argument in § 2.6 (ii).

<u>9.5</u> We now define for X_A a group analogous to $\mathbb{B}(X_v/\mathfrak{k}_v)$ in the local case (cf. § 7.8). Write, for shortness,

$$\mathbb{B}(X)_v := \mathbb{B}(X_v/\mathfrak{k}_v), \text{ for every place v of } \mathfrak{k}, \tag{5-1}$$

$$\mathbb{B}(X)_v^\circ := \mathbb{B}(X_v, X_v^\circ/\mathfrak{k}_v) \text{ for every finite place v (cf. § 7.9),} \tag{5-2}$$

so $\mathbb{B}(X)_v^\circ$ is a subgroup of $\mathbb{B}(X)_v$. Let for each place v an operator S_v in $\mathbb{B}(X)_v$ be given and suppose that S_v is in $\mathbb{B}(X)_v^\circ$ for almost all v. Then there is a unique automorphism of $L^2(X_A)$, the <u>tensor product</u> of the operators S_v, denoted by

$$S := \bigotimes_v S_v \in \mathrm{Aut}(L^2(X_A)) \text{ ,} \tag{5-3}$$

such that for $\Phi^\otimes$ of the form (4-7b)

$$S\Phi^{\otimes} = \underset{v}{\otimes} S_v\Phi_v$$

(note that $S_v\Phi_v = \varphi_v^\circ$ for almost all v [§ 4.9, Remark 3]). Indeed, it is readily seen that $S\Phi^{\otimes}$ is uniquely defined (the representation (4-7) is not unique!), and that it extends uniquely, by additivity, to the linear subspace $\mathfrak{S}(X_A)$ of all (finite) sums of functions (4-7b), yielding an automorphism of $\mathfrak{S}(X_A)$ as a subspace of the Hilbert space $L^2(X_A)$. $\mathfrak{S}(X_A)$ is dense in $\mathfrak{S}^1(X_A)$, hence in $L^2(X_A)$. Thus we obtain a unique extension S, by continuity, to an automorphism of $L^2(X_A)$. This is entirely analogous to the 'finite' tensor product (4-2); cf. also § 9.7 below.

<u>The operators of the form</u> (5-3), <u>where</u> $S_v \in \mathbb{B}(X)_v$ <u>for all</u> v, $S_v \in \mathbb{B}(X)_v^\circ$ <u>for almost all</u> v, <u>form a subgroup of</u> $\mathrm{Aut}(L^2(X_A))$ <u>which we denote by</u> $\mathbb{B}(X)_A$. This is a topological group in the strong operator topology. It does not depend on the choice of the basis X° (cf. § 9.1, Remark). We shall see later that it plays for X_A the same role as the local groups (5-1) do for X_v.

<u>9.6</u> There is a simple relation between $\mathfrak{S}^1(X_P)$ and $\mathfrak{S}^1(X_A)$. <u>Let</u> Φ_P <u>be in</u> $\mathfrak{S}^1(X_P)$ <u>and define</u>

$$\Phi := i_P(\Phi_P) := \Phi_P \otimes \underset{v \notin P}{\otimes} \varphi_v^\circ, \text{ i.e. } \Phi(x) := \Phi_P((x_v)_{v\in P}) \cdot \prod_{v\notin P} \varphi_v^\circ(x_v) ,$$

where φ_v° is the characteristic function of X_v°. Then i_P <u>is an isometric embedding of</u> $\mathfrak{S}^1(X_P)$ <u>into</u> $\mathfrak{S}^1(X_A)$: $i_P(\Phi_P)$ <u>is in</u> $\mathfrak{S}^1(X_A)$ <u>and</u>

$$\|i_P(\Phi_P)\|_{(A)} = \|\Phi_P\|_{(P)} . \tag{6-1}$$

<u>Also, if</u> (4-1) <u>is a tensor representation of</u> Φ <u>in</u> $\mathfrak{S}^1(X_P)$, <u>then</u>

$$i_P(\Phi_P) = \sum_{n\geq 1} i_P(\Phi_{Pn}^{\otimes}) \tag{6-2}$$

<u>is a tensor representation of</u> $\Phi := i_P(\Phi_P)$ <u>in</u> $\mathfrak{S}^1(X_A)$.

We shall use the notation

$$\mathfrak{S}_P(X_A) := i_P(\mathfrak{S}^1(X_P)) . \tag{6-3}$$

For the proof let us put

$$\Phi := i_P(\Phi_P),\ \Phi_n^{\otimes} := i_P(\Phi_{Pn}^{\otimes}).$$

Then we have:

$$\Phi = \sum_{n\geq 1} \Phi_n^{\otimes}, \text{ where } \Phi_n^{\otimes} := \underset{v}{\otimes} \Phi_{vn}, \text{ with } \Phi_{vn} \text{ as in (4-1b) for } v\in P \text{ and } \Phi_{vn} = \varphi_v^\circ \text{ for } v \notin P.$$

Since $\|\varphi_v^\circ\|_{(v)} = 1$, we also have

$$\sum_{n\geq 1} \prod_{v} \|\Phi_{vn}\|_{(v)} = \sum_{n\geq 1} \prod_{v\in P} \|\Phi_{vn}\|_{(v)} < \infty ,$$

hence Φ is in $\mathfrak{S}^1(X_A)$ and relation (6-2) holds. We conclude that

$$\|\Phi\|_{(A)} \le \|\Phi_P\|_{(P)}. \tag{6-4}$$

To show the reverse inequality, consider an arbitrary tensor representation (4-8) of Φ. Let $x \in X_A$ be of the form $x_P x (a_v)_{v \notin P}$, where $x_P \in X_P$ is variable and $a_v \in X_v^\circ$, $v \notin P$, is fixed, say $a_v = 0$. Then (4-8) yields, for such x,

$$\Phi_P(x_P) = \sum_{n\ge1} c_n \Phi_{Pn}(x_P),$$

where

$$\Phi_{Pn} := \underset{v\in P}{\otimes} \Phi_{vn}, \quad c_n := \prod_{v\notin P} \Phi_{vn}(a_v), \text{ with } \Phi_{vn} \text{ as in (4-8b,c).}$$

Now by (4-6) we have

$$|c_n| \le \prod_{v\notin P} \|\Phi_{vn}\|_{(v)} .$$

Choose any $v_1 \in P$ and put

$$\Phi'_{v_1 n} := c_n \Phi_{v_1 n}, \quad \Phi'_{vn} := \Phi_{vn} \text{ if } v \in P,\ v \neq v_1 .$$

Then

$$\Phi_P = \sum_{n\ge1} \Phi^{\otimes}_{Pn}, \text{ with } \Phi^{\otimes}_{Pn} := \underset{v\in P}{\otimes} \Phi'_{vn}, \sum_{n\ge1} \prod_{v\in P} \|\Phi'_{vn}\|_{(v)} \le \sum_{n\ge1} \prod_{v} \|\Phi_{vn}\|_{(v)},$$

and this tensor representation in $\mathfrak{S}^1(X_P)$ yields the reverse of (6-4); thus (6-1) holds.

<u>9.7</u> LEMMA. <u>Let</u> S <u>be in</u> $\mathbb{B}(X)_A$. <u>Suppose</u> $S = \underset{v}{\otimes} S_v$ <u>with</u> $S_v \in \mathbb{B}(X)_v^\circ$ <u>for</u> $v \notin P$, <u>say, and define</u> $S_P := \underset{v\in P}{\otimes} S_v \in \mathrm{Aut}(L^2(X_P))$. <u>Let</u> Φ_P <u>be in</u> $\mathfrak{S}^1(X_P)$ <u>so that</u> $i_P(\Phi_P)$ <u>is in</u> $\mathfrak{S}^1(X_A)$. <u>Then</u> $Si_P(\Phi_P)$ <u>is in</u> $\mathfrak{S}^1(X_A)$ <u>and</u>

$$Si_P(\Phi_P) = i_P(S_P\Phi_P). \tag{7-1}$$

Let Φ_P have the tensor representation (4-1). Then $S_P\Phi_P \in \mathfrak{S}^1(X_P)$ has the tensor representation (4-4). We can now use (6-2), and apply it also to $S_P\Phi_P$. This shows that

$$i_P(\Phi_P) = \sum_{n\ge1} i_P(\Phi^{\otimes}_{Pn}), \quad i_P(S_P\Phi_P) = \sum_{n\ge1} i_P(S_P\Phi^{\otimes}_{Pn}) \tag{7-2}$$

are tensor representations of $i_P(\Phi_P)$ and $i_P(S_P\Phi_P)$ in $\mathfrak{S}^1(X_A)$. For each $N \ge 1$ we have

$$S \sum_{n=1}^{N} i_P(\Phi^{\otimes}_{Pn}) = \sum_{n=1}^{N} i_P(S_P\Phi^{\otimes}_{Pn}),$$

and for $N \longrightarrow \infty$ we obtain, since both series in (7-2) converge in $L^2(X_A)$ [§ 2.3, Remark 2],

$$Si_P(\Phi_P) = \sum_{n\ge1} i_P(S_P\Phi^{\otimes}_{Pn}). \tag{7-3}$$

Comparison of (7-2) and (7-3) yields (7-1).

COROLLARY. _If_ S _is as in the lemma above, then_ $\Phi \longmapsto S\Phi$, $\Phi \in \mathfrak{S}_P(X_A)$, _is an automorphism of_ $\mathfrak{S}_P(X_A)$, as defined in (6-3).

This follows at once if we combine the invariance of φ_v°, $v \notin P$, under $S_v \in \mathbb{B}(X)_v^\circ$ [§ 4.9, Remark 3] with (4-1) and the lemma above.

_9.8_ The argument of § 9.6 will also show this. Let

$$i_{P,P'}(\Phi_P) := \Phi_P \otimes \underset{\substack{v \in P' \\ v \notin P}}{\otimes} \varphi_v^\circ \ , \quad \Phi_P \in \mathfrak{S}^1(X_P),\ P' \supset P \supset P_\infty,\ P' \neq P.$$

Then $i_{P,P'}$ _is an isometric embedding of_ $\mathfrak{S}^1(X_P)$ _into_ $\mathfrak{S}^1(X_{P'})$. _Also_

$$i_P = i_{P'} \circ i_{P,P'} \ , \quad i_{P,P''} = i_{P'P''} \circ i_{P,P'} \ , \quad P'' \supset P' \supset P \supset P_\infty \, .$$

A tensor representation (4-1) of Φ_P in $\mathfrak{S}^1(X_P)$ gives rise to a tensor representation of $i_{P,P'}(\Phi_P)$ in $\mathfrak{S}^1(X_{P'})$ in the obvious way.

_9.9_ We now introduce a certain subgroup $B_0(X)_A$ of $B_0(X_A)$ which will be the analogue of the group $B_0(X_v/\mathfrak{k}_v)$ defined in the local case. We write, for shortness (cf. §§ 7.8, 7.9):

$$B_0(X)_v := B_0(X_v/\mathfrak{k}_v); \ B_0(X)_v^\circ := B_0(X_v, X_v^\circ/\mathfrak{k}_v) \text{ for finite } v. \qquad (9\text{-}1)$$

$B_0(X)_v^\circ$ is a subgroup of $B_0(X)_v$. We consider the 'sequences' $(\sigma_v, \psi_v)_v$ such that

(i) (σ_v, ψ_v) is in $B_0(X)_v$ for all v,
(ii) (σ_v, ψ_v) is in $B_0(X)_v^\circ$ for almost all v.

For such a sequence, $\underset{v}{\otimes}\, \psi_v$ is well-defined, and $(\sigma_v)_v$ is an element of $Sp(X)_A$ [§ 9.3].

We define _the group_ $B_0(X)_A$ as follows. $B_0(X)_A$ consists of all pairs $((\sigma_v)_v, \underset{v}{\otimes}\, \psi_v)$ such that the sequence $(\sigma_v, \psi_v)_v$ has the properties (i) and (ii) above. The multiplication is that resulting from the local groups $B_0(X)_v$. $B_0(X)_A$ does not depend on the choice of the basis X°.

We now exhibit $B_0(X)_A$ as a subgroup of $B_0(X_A)$ and show its relation to the group $\mathbb{B}(X)_A$ [§ 9.5]. Let $a = (a_v)_v \in X_A$, $a^* = (a_v^*)_v \in X_A^*$, and let the operators $U(a,a^*)$ in $L^2(X_A)$ and $U_v(a_v, a_v^*)$ in $L^2(X_v)$ be defined as usual [§ 1.8]. We have [cf. (5-3)]:

$$U(a,a^*) = \underset{v}{\otimes}\, U_v(a_v, a_v^*) \in \mathbb{B}(X)_A .$$

Thus, if $S = \underset{v}{\otimes}\, S_v \in \mathbb{B}(X)_A$, then

$$S^{-1}U(a,a^*)S = \underset{v}{\otimes}\, S_v^{-1}U_v(a_v,a_v^*)S_v = \underset{v}{\otimes}\, [\psi_v(a_v,a_v^*)U_v((a_v,a_v^*)\sigma_v)] =$$
$$= \psi(a,a^*) \cdot \underset{v}{\otimes}\, U_v((a_v,a_v^*)\sigma_v), \text{ with } \psi := \underset{v}{\otimes}\, \psi_v,$$

or, in the usual notation: <u>for</u> $\underset{v}{\otimes} S_v \in \mathbb{B}(X)_A$ <u>the relation</u>

$$\pi_o(\underset{v}{\otimes} S_v) = ((\sigma_v)_v, \underset{v}{\otimes} \psi_v) \text{ if } \pi_{ov}(S_v) = (\sigma_v, \psi_v) \in B_o(X)_v \tag{9-2}$$

<u>holds</u>. Thus $\mathbb{B}(X)_A$ <u>is a subgroup of</u> $\mathbb{B}(X_A)$.

Now let $((\sigma_v)_v, \underset{v}{\otimes} \psi_v)$ in $B_o(X)_A$ be given. Then § 7.8 shows that for each v there is an S_v in $\mathbb{B}(X)_v$ such that $\pi_{ov}(S_v) = (\sigma_v, \psi_v)$, and when (σ_v, ψ_v) is in $B_o(X)_v^\circ$, we can take S_v in $\mathbb{B}(X)_v^\circ$ (cf. § 7.9). Thus, with the usual notation,

$$\pi_o(\mathbb{B}(X)_A) = B_o(X)_A.$$

This also shows that $B_o(X)_A$ <u>lies in</u> $B_o(X_A)$. Obviously, $\mathbb{B}(X)_A$ <u>contains the kernel</u> $T \cdot I$ <u>of</u> π_o. We have proved:

PROPOSITION. <u>The groups</u> $\mathbb{B}(X)_A$ <u>and</u> $B_o(X)_A$ <u>are contained in</u> $\mathbb{B}(X_A)$, $B_o(X_A)$, <u>respectively, and</u> $\pi_o(\mathbb{B}(X)_A) = B_o(X)_A$. <u>Moreover</u>, $\mathbb{B}(X)_A$ <u>is the subgroup of all</u> S <u>in</u> $\mathbb{B}(X_A)$ <u>such that</u> $\pi_o(S)$ <u>lies in</u> $B_o(X)_A$.

REMARK. We can introduce a topology in $B_o(X)_A$ as a subgroup of $B_o(X_A)$ [§ 4.10]. <u>The mapping</u> $((\sigma_v, \psi_v))_v \longmapsto (\sigma_v)_v$ <u>is then a strict morphism of</u> $B_o(X)_A$ <u>onto</u> $Sp(X)_A$, and an isomorphism if char $\mathfrak{k} \neq 2$ [cf. § 7.9, Remark].

<u>9.10</u> Consider the adelic pseudosymplectic group $Ps(X)_A$ [§ 9.3]. There is a natural homomorphism μ_A of $Ps(X)_A$ into $B_o(X)_A$, analogous to μ, as defined in § 8.1 for the local case. Let $\chi = \underset{v}{\otimes} \chi_v$ be a non-trivial character of $\mathfrak{k}_A$ trivial on $\mathfrak{k}$ and recall that, if a basis X° of X is chosen, then $(X_v^\circ)_* = (X^*)_v^\circ$ for almost all finite v, more precisely for $v \notin P_o$ [(2-2), (2-3)]. It follows that for each (σ_v, f_v) in $Ps(X)_v^\circ$ we have: $\mu_v(\sigma_v, f_v) := (\sigma_v, \chi_v \circ f_v)$ <u>is in</u> $B_o(X)_v^\circ$ <u>for</u> $v \notin P_o$, or

$$\mu_v(Ps(X)_v^\circ) \subset B_o(X)_v^\circ \quad \text{[cf. (9-1)]}, \ v \notin P_o. \tag{10-1}$$

Hence we can define for $s = ((\sigma_v, f_v))_v \in Ps(X)_A$

$$\mu_A(s) = \mu_A(((\sigma_v, f_v))_v) := ((\sigma_v)_v, \underset{v}{\otimes}(\chi_v \circ f_v)) \in B_o(X)_A, \tag{10-2}$$

and μ_A <u>is a homomorphism of</u> $Ps(X)_A$ <u>into</u> $B_o(X)_A$.

<u>If</u> char $\mathfrak{k} \neq 2$, <u>then</u> μ_A <u>is injective</u>, in analogy with the local case [§ 8.1]: the kernel of μ_A consists of all $s = ((\sigma_v, f_v))_v$ in $Ps(X)_A$ such that $\sigma_v = 1_v \in Sp(X)_v$ for all v and $\underset{v}{\otimes}(\chi_v \circ f_v) = 1$; but the first condition already implies that f_v is additive, for each v, so $f_v = 0$. In this case we also have $Ps(X)_A \cong Sp(X)_A$, as remarked at the end of § 9.3. Thus, <u>if</u> char $\mathfrak{k} \neq 2$, <u>we may identify</u> $\mu_A(Ps(X)_A)$ <u>with</u> $Sp(X)_A$.

We can now introduce the <u>metaplectic group</u> $Mp(X)_A$ belonging to X_A: we define $Mp(X)_A$ as the subgroup of the product $Ps(X)_A \times \mathbb{B}(X)_A$ formed by all pairs (s,S) such that $\mu_A(s) = \pi_o(S)$. Thus $Mp(X)_A$ is a topological group, with the topology induced by the product topology (it is the 'fibre product' of $Ps(X)_A$ and $\mathbb{B}(X)_A$ over $B_o(X_A)$). Since $\mu_A(Ps(X)_A)$ lies in $B_o(X)_A$, the proposition in § 9.9 shows that one can also define $Mp(X)_A$ as the fibre product of $Ps(X)_A$ and $\mathbb{B}(X_A)$ over $B_o(X_A)$; this definition is given by Weil [25, n° 37, p. 188].

We write again [cf. § 8.4]

$$\pi(\underline{S}) := s \ , \text{ if } \underline{S} := (s,S) \in Mp(X)_A \ , \tag{10-3}$$

so π is a morphism of $Mp(X)_A$ onto $Ps(X)_A$ (cf. § 9.9, Proposition). Clearly π is continuous; we will show below that π is also strict (i.e. open). Let us put

$$Mp(X)_P^\circ := \pi^{-1}(Ps(X)_P^\circ), \quad P \supset P_\infty \text{ finite.} \tag{10-4}$$

$Mp(X)_P^\circ$ is an open subgroup of $Mp(X)_A$. Thus we can write

$$Mp(X)_A = \operatorname{ind.lim}_{P\supset P_\infty} Mp(X)_P^\circ .$$

The subgroups $Mp(X)_P^\circ$ will play an important role in the sequel.

9.11 We next show that $Mp(X)_A$ is locally compact and π is open. The method is quite analogous to that in the local case [§ 8.5].

For each place v of $\mathfrak{k}$ we have an open set Ω_v in $Ps(X)_v$ and a continuous lifting R_v of Ω_v to $\mathbb{B}(X)_v$:

$$s_v \longmapsto R_v(s_v) := R_{ov}(\mu_v(s_v)), \quad s_v \in \Omega_v. \tag{11-1}$$

We also have a second lifting method for almost all finite v. Let P_o be as before [(2-2), (2-3), (10-1)]. Then we can apply the theta lifting described in § 4.9 [cf. also § 7.9] to $B_o(X)_v^\circ$. Let us denote this lifting by R'_{ov}. We can now introduce the following lifting R'_v of $Ps(X)_v^\circ$ into $\mathbb{B}(X)_v$:

$$s_v \longmapsto R'_v(s_v) := R'_{ov}(\mu_v(s_v)), \quad s_v \in Ps(X)_v^\circ \quad (v \notin P_o). \tag{11-2}$$

In view of (10-1), we have

$$R'_v(Ps(X)_v^\circ) \subset \mathbb{B}(X)_v^\circ, \quad v \notin P_o \ [\text{cf. } (5\text{-}2)].$$

Further, let

$$Mp(X)_v^\circ := \left\{ (s_v, R'_v(s_v) \mid s_v \in Ps(X)_v^\circ \right\}, \ v \notin P_o. \tag{11-3}$$

This is clearly a subgroup of

$$Mp(X)_v := Mp(X_v).$$

We shall show below that the mapping (11-2) of $Ps(X)_v^\circ$ into $Aut(L^2(X_v))$ is continuous. Then it will follow [§ 1.12] that $Mp(X)_v^\circ$ is isomorphic, as a topological group, <u>to</u> $Ps(X)^\circ_v$, <u>thus compact</u>.

Let us note that for $P \supset P_0$

$$Mp(X)^\circ_P = \{((s_v)_v, \otimes_v S_v) \mid (s_v, S_v) \in Mp(X)_v, v \in P, (s_v, S_v) \in Mp(X)^\circ_v, v \notin P\}. \tag{11-4}$$

To prove the continuity of (11-2), we refer to § 4.9 and replace, in the notation introduced there, G by X_v, Γ by X°_v. So we are dealing with theta functions in $L^2(X_v, X^\circ_v)$. We take $v \notin P_0$ [(2-2), (2-3), (10-1)] and will show: for θ in $\mathcal{K}(X_v, X^\circ_v)$ the mapping of $Ps(X)^\circ_v$ into $\mathcal{K}(X_v, X^\circ_v)$ defined in § 4.9 by

$$s_v = (\sigma_v, f_v) \longmapsto \tilde{R}_v(s_v)\theta, \ [\tilde{R}_v(s_v)\theta](z_v) := \theta(z_v\sigma_v)\cdot\chi_v \circ f_v(z_v), \quad s_v \in Ps(X)^\circ_v,$$

is continuous, $\mathcal{K}(X_v, X^\circ_v)$ being provided with the norm in $L^2(X_v, X^\circ_v)$. Now θ vanishes on $X_v \times X^*_v$ outside some compact set $K_c := c\cdot(X^\circ_v \times (X^*)^\circ_v)$ [cf. (2-3)] with suitable $c = c_\theta$ in $\mathfrak{f}_v$. Moreover, K_c is invariant under σ_v for $(\sigma_v, f_v) \in Ps(X)^\circ_v$. Further, the subspace $\mathcal{K}_c(X_v, X^\circ_v)$ of all θ in $\mathcal{K}(X_v, X^\circ_v)$ with support in K_c is invariant under $\tilde{R}_v(s_v)$, $s_v \in Ps(X)^\circ_v$. The mapping above is continuous if we consider only the subspace $\mathcal{K}_c(X_v, X^\circ_v)$; this is seen by introducing the supremum norm there. It follows that the mapping (11-2) is continuous in the strong topology of $Aut(L^2(X_v))$.

<u>9.12</u> We now put $\Omega_v := \Omega(X_v)$ and

$$\Omega^\circ_P := \prod_{v \in P} \Omega_v \times \prod_{v \notin P} Ps(X)^\circ_v, \quad P \supset P_0.$$

We then define a lifting R°_P of Ω°_P to $\mathbb{B}(X)_A$,

$$s \longmapsto R^\circ_P(s) \in \mathbb{B}(X)_A, \ s = (s_v)_v \in \Omega^\circ_P, \ P \supset P_0, \tag{12-1}$$

by putting

$$R^\circ_P(s) := \otimes_v S_v, \ S_v := \begin{cases} R_{0v}(s_v), & v \in P \text{ [cf. (11-1)]}, \\ R'_{0v}(s_v), & v \notin P \text{ [cf. (11-2)]}. \end{cases}$$

It will be shown below that <u>the lifting</u> (12-1) <u>is continuous</u>. The relation (9-2) and the definition of μ_A [(10-2)] show that, with the usual notation,

$$\pi_0(R^\circ_P(s)) = \mu_A(s), \ s \in \Omega^\circ_P,$$

thus $(s, R^\circ_P(s))$ is in $Mp(X)_A$ and

$$s \longmapsto (s, R^\circ_P(s)), \ s \in \Omega^\circ_P, \ P \supset P_0, \tag{12-2}$$

<u>is a continuous lifting of</u> Ω°_P <u>to</u> $Mp(X)_A$, in fact to $Mp(X)^\circ_P$. We then have, using again §§ 1.12, 1.13, as in the local case [§ 8.5]: (i) <u>the</u>

<u>mapping</u> (12-2) <u>is a homeomorphism of</u> Ω_P° <u>into</u> $Mp(X)_P^\circ$; (ii) <u>the mapping</u>

$$(\tau,s) \longmapsto (s,\tau\cdot R_P^\circ(s)) \ , \ (\tau,s) \in T\times\Omega_P^\circ \ ,$$

<u>is a homeomorphism of</u> $T\times\Omega_P^\circ$ <u>onto</u> $\pi^{-1}(\Omega_P^\circ)$, which is an open set in $Mp(X)_P^\circ$, thus in $Mp(X)_A$. Hence $Mp(X)_A$ <u>is locally compact and the morphism</u> π [(10-3)] <u>is strict</u>, i.e. not only continuous, but also open.

It remains to show that <u>the lifting</u> (12-1) <u>is continuous</u>. This amounts to showing that for $\Phi\in L^2(X_A)$ the mapping $s \longmapsto R_P^\circ(s)\Phi$ of Ω_P° into $L^2(X_A)$ is continuous. It is enough to take Φ in $\mathcal{S}(X_A)$ [§ 9.5] which is dense in $L^2(X_A)$. Thus, in fact, we need only consider functions $\Phi^\otimes$ as in (4-7b). Suppose, for such a $\Phi^\otimes$, that $\Phi_v = \varphi_v^\circ$ for all $v \notin P_1$, and put $P' := P\cup P_1$ $(P \supset P_o)$. Then we can write

$$\Phi^\otimes = \otimes_{v\in P'}\Phi_v \otimes \otimes_{v\notin P'}\varphi_v^\circ \ , \tag{12-3a}$$

$$R_P^\circ(s)\Phi^\otimes = \otimes_{v\in P'}\Phi'_v \otimes \otimes_{v\notin P'}\varphi_v^\circ, \ s\in\Omega_P^\circ \ , \tag{12-3b}$$

where

$$\Phi'_v := R_v(s_v)\Phi_v, \ v\in P, \ \Phi'_v := R'_v(s_v)\Phi_v \ , \ v\notin P, \ v\in P' .$$

The right-hand term in (12-3b) depends continuously, in $L^2(X_A)$, on $\Phi'_v \in L^2(X_v)$, $v\in P'$. Thus the asserted continuity follows from the continuity, in $L^2(X_v)$, of each local mapping $s_v \longmapsto R_v(s_v)\Phi_v$, $s_v\in\Omega_v$ $(v\in P)$ [§ 8.5] and $s_v \longmapsto R'_v(s_v)\Phi_v$ $(v\notin P, \ v\in P')$ [§ 9.11].

<u>9.13</u> The topology of $Mp(X)_A$ can be described in a simple way which will be useful for later purposes. We recall (11-3) and (11-4).

LEMMA. <u>Let, for</u> $P \supset P_o \supset P_\infty$ [(2-3),(10-1)], U_v <u>be a neighbourhood of the neutral element of</u> $Mp(X)_v$, <u>for each</u> $v\in P$. <u>Then</u>

$$\left\{((s_v)_v, \otimes_v S_v) \mid (s_v,S_v)\in U_v, \ v\in P, \ (s_v,S_v)\in Mp(X)_v^\circ, \ v\notin P\right\} \quad P\supset P_o$$

<u>is a neighbourhood of the neutral element in</u> $Mp(X)_P^\circ$, <u>hence in</u> $Mp(X)_A$. <u>Moreover, the neighbourhoods of this type form a fundamental system of neighbourhoods of the neutral element in</u> $Mp(X)_A$.

This lemma may also be expressed as follows if we introduce the group

$$M_*(X)_P^\circ := \prod_{v\in P} Mp(X)_v \times \prod_{v\notin P} Mp(X)_v^\circ \ , \ P\supset P_o \ ,$$

<u>The mapping of</u> $M_*(X)_P^\circ$ <u>onto</u> $Mp(X)_P^\circ$ <u>defined by</u>

$$((s_v,S_v))_v \longmapsto ((s_v)_v, \otimes_v S_v), \ ((s_v,S_v))_v \in M_*(X)_P^\circ, \tag{13-1}$$

<u>is a strict morphism</u> (cf. Weil [25, n° 45, p.199]).

This mapping is clearly an algebraic homomorphism. To show that it is continuous and open, define the open subset Ω_* of $M_*(X)^\circ_P$ by

$$\Omega_* := \prod_{v\in P} \pi_v^{-1}(\Omega_v)\times \prod_{v\notin P} Mp(X)^\circ_v .$$

Thus Ω_* consists of all sequences $((s_v,S_v))_v$ such that

if $v\in P$, then $s_v\in \Omega_v$, $S_v = \tau_v\cdot R_v(s_v)$, $\tau_v\in T$,

if $v \notin P$, then $s_v\in Ps(X)^\circ_v$, $S_v = R'_v(s_v)$.

By (13-1) $((s_v,S_v))_v\in \Omega_*$ is mapped onto $((s_v)_v,\tau\cdot\bigotimes_v S_v)\in Mp(X)_A$, where

$$\tau := \prod_{v\in P}\tau_v . \tag{13-2}$$

That the restriction of (13-1) to Ω_* is continuous and open, is now clear if we split it into three consecutive mappings:

(i) the natural homeomorphism of Ω_* onto

$$T^{\mathrm{card}\ P}\times \prod_{v\in P}\Omega_v\times \prod_{v\notin P} Ps(X)^\circ_v = T^{\mathrm{card}\ P}\times\Omega^\circ_P ,$$

(ii) the mapping of $T^{\mathrm{card}\ P}\times\Omega^\circ_P$ onto $T\times\Omega^\circ_P$ derived from (13-2). This is obviously continuous and open,

(iii) the homeomorphism of $T\times\Omega^\circ_P$ onto $\pi^{-1}(\Omega^\circ_P)$.

This shows indeed that (13-1) is a strict morphism.

<u>9.14</u> We need a certain subspace of $\mathfrak{S}^1(X_A)$ to formulate the main result for the metaplectic group in the adelic case. Let $\mathfrak{S}_I(X_A)$ be the linear subspace of $\mathfrak{S}^1(X_A)$ defined by

$$\mathfrak{S}_I(X_A) := \bigcup_{P\supset P_\infty} \mathfrak{S}_P(X_A),$$

where $\mathfrak{S}_P(X_A) := i_P(\mathfrak{S}^1(X_P))$ [cf. (6-3)]. We provide $\mathfrak{S}_I(X_A)$ with the inductive limit topology of the norm topologies of the Banach spaces $\mathfrak{S}_P(X_A)$. This is a finer topology than that induced in $\mathfrak{S}_I(X_A)$ by the norm of $\mathfrak{S}^1(X_A)$: V is a nd. of 0 in $\mathfrak{S}_I(X_A)$ if, and only if, $V\cap\mathfrak{S}_P(X_A)$ is a nd. of 0 in $\mathfrak{S}_P(X_A)$ for every $P \supset P_\infty$. Thus

$$\mathfrak{S}_I(X_A) := \operatorname{ind.lim}_{P\supset P_\infty} \mathfrak{S}_P(X_A),$$

and in view of § 9.8 we may replace here $\mathfrak{S}_P(X_A)$ by $\mathfrak{S}^1(X_P)$. It is a strict inductive limit in a slightly generalized sense, the sets P being finite subsets of a countable set.

Thus $\mathfrak{S}_I(X_A)$ is a separated locally convex vector space over $\mathbb{C}$ which is complete, but not metrizable (for this, and various other properties cf. e.g. Bourbaki [2, Ch. II, §-4, n°s 4, 6, § 8, n° 2], Robertson [21, Ch. V, §§ 2, 3, Ch. VII, § 1]. We shall have to consider <u>linear functionals on</u> $\mathfrak{S}_I(X_A)$, <u>and linear mappings of</u> $\mathfrak{S}_I(X_A)$ <u>into itself</u>;

what is of relevance here is that these *are continuous if and only if they are continuous on each subspace* $\mathfrak{S}_P(X_A)$.

We call $\mathfrak{S}_I(X_A)$ the *inductive Segal algebra of the adele group* X_A. It may be observed that the topology of $\mathfrak{S}_I(X_A)$ is quite analogous to that of X_A. Moreover, $\mathfrak{S}_I(X_A)$ contains $\mathcal{S}(X_A)$: cf. the description of $\mathcal{S}(X_A)$ given by Weil in [25, n°s 29, 39]. An application of the general result in § 2.14 to each group X_v shows that $\mathcal{S}(X_A)$ is a small - but of course dense - part of $\mathfrak{S}_I(X_A)$.

9.15 We can now state and prove the main result about the Segal representation of $Mp(X)_A$ in the adelic case.

THEOREM. *The unitary representation* $\underline{S} = (s,S) \longmapsto S$ *of the metaplectic group* $Mp(X)_A$ *in* $L^2(X_A)$ *induces a continuous representation of* $Mp(X_A)$ *in the topological vector space* $\mathfrak{S}_I(X_A)$. *Moreover, for* P *large, it induces a continuous representation of the open subgroup* $Mp(X)^\circ_P$ *in the Banach space* $\mathfrak{S}_P(X_A)$ (more precisely for $P \supset P_o$, where P_o is defined as in § 9.2 [cf. also (10-1)]).

Explicitly:

Let $\underline{S} = (s,S) \in Mp(X_A)$, $\Phi \in \mathfrak{S}_I(X_A)$, $\underline{S}\Phi := S\Phi$ (Weil's notation). Then $\Phi \longmapsto \underline{S}\Phi$ is an automorphism of the inductive Segal algebra $\mathfrak{S}_I(X_A)$, as a topological vector space, and the mapping $\underline{S} \longmapsto \underline{S}\Phi$ of $Mp(X_A)$ into $\mathfrak{S}_I(X_A)$ is continuous. Moreover, if $P \supset P_o$, then for $\underline{S} \in Mp(X)^\circ_P$ the mapping $\Phi \longmapsto \underline{S}\Phi$, $\Phi \in \mathfrak{S}_P(X_A)$, is an automorphism of $\mathfrak{S}_P(X_A)$, and for fixed $\Phi \in \mathfrak{S}_P(X_A)$ the mapping $\underline{S} \longmapsto \underline{S}\Phi$ of $Mp(X)^\circ_P$ into $\mathfrak{S}_P(X_A)$ is continuous.

The corollary in § 9.7 implies that $\underline{S} \in Mp(X)^\circ_P$ defines for $P \supset P_o$ an automorphism of $\mathfrak{S}_P(X_A)$. Now, for any $P' \supset P$ we have $Mp(X)^\circ_P \subset Mp(X)^\circ_{P'}$, hence $\underline{S} \in Mp(X)^\circ_P$ also defines an automorphism of $\mathfrak{S}_{P'}(X_A)$. Thus $\underline{S}$ defines an automorphism of $\mathfrak{S}_I(X_A)$ (which has the inductive limit topology). Also, $\underline{S}$ can be an arbitrary element of $Mp(X)_A$, since every $\underline{S}$ in $Mp(X)_A$ lies in $Mp(X)^\circ_P$ for P sufficiently large [cf. (10-4)]. The automorphism part of the theorem is thus simply a combination of the corollary in § 9.7 with various properties of $Mp(X)_A$ and $\mathfrak{S}_I(X_A)$.

The proof of the other part of the theorem, concerning the continuity of the representation, depends on the corresponding local result in § 8.6 and on the proposition in § 8.7, together with a number of lemmas proved previously.

Consider first $Mp(X)^\circ_P$: it leaves $\mathfrak{S}_P(X_A)$ invariant if $P \supset P_o$. More precisely, if $\Phi = i_P(\Phi_P)$, $\Phi_P \in \mathfrak{S}^1(X_P)$, then [cf. § 9.7, Lemma] $\underline{S}\Phi =$

$= i_P(S_P\Phi_P)$ for $\underline{S} = (s,S) \in Mp(X)^\circ_P$ and [cf. (6-1)]

$$\|S\Phi - \Phi\|_{(A)} = \|S_P\Phi_P - \Phi_P\|_{(P)} .$$

This relation will yield continuity at the neutral element which will be sufficient. We have to prove the following:

LEMMA. Let Φ_P be in $\mathfrak{S}^1(X_P)$. Given $\varepsilon > 0$, there are, for each $v \in P$, neighbourhoods U_v of e_v in $Mp(X)_v$ such that if (s_v, S_v) is in U_v, then $S_P := \bigotimes_{v \in P} S_v$ satisfies

$$\|S_P\Phi_P - \Phi_P\|_{(P)} < \varepsilon .$$

This lemma means that the representation $(\underline{S}_v)_{v\in P} \longmapsto S_P\Phi_P$ of $\prod_{v\in P} Mp(X)_v$ in $\mathfrak{S}^1(X_P)$ is continuous. The proof consists of several steps.

(i) First we have: Let Φ_v, $v \in P$, be in $\mathfrak{S}^1(X_v)$. The mapping

$$(\Phi_v)_{v\in P} \longmapsto \bigotimes_{v\in P} \Phi_v$$

of $\prod_{v\in P} \mathfrak{S}^1(X_v)$ into $\mathfrak{S}^1(X_P)$ is continuous. This was proved in § 2.7.

(ii) Combining (i) with the local theorem in § 8.6, we obtain immediately, by the very definition of S_P above: for Φ_P of the form $\Phi_P = \Phi_P^{\otimes} := \bigotimes_{v\in P}\Phi_v$ the lemma holds. Hence the lemma also holds if Φ_P is in $\mathfrak{S}(X_P)$, i.e. a (finite) sum of functions $\Phi_P^{\otimes}$.

(iii) We now apply the proposition in § 8.7. Let U'_v be a compact nd. of the neutral element in $Mp(X)_v$, for each $v \in P$. Then there are constants $C_v = C_v(U'_v)$ such that

$$\|S_v\Phi_v\|_{(v)} \le C_v \cdot \|\Phi_v\|_{(v)} \quad \text{for all } \Phi_v \in \mathfrak{S}^1(X_v) \text{ if } (s_v, S_v) \in U'_v \ (v \in P).$$

Put $C := \prod_{v\in P} C_v$. Then [cf. (4-5)]

$$\|S_P\Phi_P\|_{(P)} \le C \cdot \|\Phi_P\|_{(P)} \quad \text{for all } \Phi_P \in \mathfrak{S}^1(X_P) \text{ if } (s_v, S_v) \in U'_v \text{ for all } v \in P. \tag{15-1}$$

(iv) From (ii) and (iii) we now obtain the lemma in the general case, i.e. for any $\Phi_P \in \mathfrak{S}^1(X_P)$. Let $\Phi_P = \sum_{n\ge1} \Phi_{Pn}^{\otimes}$ be a tensor representation of Φ_P in $\mathfrak{S}^1(X_P)$ and write

$$\Phi_P = \Phi'_P + \Phi''_P, \text{ where } \Phi' := \sum_{n=1}^{N} \Phi_{Pn}^{\otimes} , \quad \Phi'' := \sum_{n>N} \Phi_{Pn}^{\otimes} ,$$

with N to be chosen. Let U'_v $(v \in P)$ and C be as in (iii) so that (15-1) holds. Then we have, independently of N,

$$\|S_P\Phi_P - \Phi_P\|_{(P)} \leq \|S_P\Phi'_P - \Phi'_P\|_{(P)} + (C+1)\cdot\|\Phi''_P\|_{(P)} = A_1+A_2, \text{ say.}$$

We can choose N so large that $A_2 < \varepsilon/2$. Then, by (ii), there are nds. U_v, $v \in P$, of e_v in $Mp(X)_v$ such that $A_1 < \varepsilon/2$ if $(s_v, S_v) \in U_v$ for all $v \in P$, and we may suppose $U_v \subset U'_v$. Thus the lemma is proved.

It now follows, in view of the lemma in § 9.13, that for fixed Φ in $\mathfrak{S}_P(X_A)$ $(P \supset P_0)$ the mapping $\underline{S} \longmapsto \underline{S}\Phi$ of $Mp(X)^\circ_P$ into $\mathfrak{S}_P(X_A)$ is continuous at the neutral element. Thus the representation of $Mp(X)^\circ_P$ in $\mathfrak{S}_P(X_A)$ is continuous.

The continuity of the representation of $Mp(X)_A$ in $\mathfrak{S}_I(X_A)$ follows from the result for $Mp(X)^\circ_P$. Indeed, in the mapping

$$\underline{S} \longmapsto \underline{S}\Phi\ ,\ \underline{S} \in Mp(X)^\circ_P \quad (\Phi \in \mathfrak{S}_P(X_A)), \tag{15-2}$$

we may replace $\mathfrak{S}_P(X_A)$ by $\mathfrak{S}_{P'}(X_A)$, with any $P' \supset P$. Hence for <u>any</u> Φ in $\mathfrak{S}_I(X_A)$ (15-2) is a continuous mapping of $Mp(X)^\circ_P$ into $\mathfrak{S}_I(X_A)$. Here we use the fact that for each P the topology of the Banach space $\mathfrak{S}_P(X_A)$ coincides with that induced by the inductive limit topology of $\mathfrak{S}_I(X_A)$. Moreover, $P \supset P_0$ was arbitrary, and continuity of the mapping (15-2) on each open subgroup $Mp(X)^\circ_P$ amounts to continuity on $Mp(X)_A$.

Thus the proof of the theorem is complete. It is an open question whether the result also holds for $\mathfrak{S}^1(X_A)$, not only for $\mathfrak{S}_I(X_A)$.

<u>9.16</u> Let us now consider the case when char $\mathfrak{k} \neq 2$ [cf. the local case, §§ 8.8, 8.9].

<u>If</u> char $\mathfrak{k} \neq 2$, <u>then the following holds</u>:

(A) $Mp(X)_A$ <u>is isomorphic to the subgroup</u> $\pi_0^{-1}(\mu_A(Ps(X)_A))$ <u>of</u> $\mathbb{B}(X)_A$.

(B) $\mathbb{B}(X)_A$ <u>is locally compact and the restriction of</u> π_0 <u>to</u> $\mathbb{B}(X)_A$ <u>is a strict morphism of</u> $\mathbb{B}(X)_A$ <u>onto</u> $B_0(X)_A$ (with the topology induced by that of $B_0(X_A)$). <u>Moreover</u>, $\mathbb{B}(X)_A$ <u>has the Segal continuity property relative to</u> $\mathfrak{S}_I(X_A)$. More precisely, for each $P \supset P_0$, the open subgroup $\mathbb{B}(X)^\circ_P$, defined in the obvious way [cf. (5-1), (5-2)], has a continuous representation in $\mathfrak{S}_P(X_A)$.

We indicate briefly the main steps of the proof. It is convenient here to change the notation used in the local case [§ 8.8]: we put, with b_1 as in § 7 (1-1),

$$b'(z) := (1/2)b_1(z,z),\ z = (x,x^*) \in X \times X^*,\ b'_\sigma(z) := b'(z\sigma) - b'(z),$$
$$\sigma \in Sp(X/\mathfrak{k}).$$

We can extend b', b'_σ to $X_v \times X^*_v$, $Sp(X)_v$, for each v. Then we obtain a <u>continuous</u> function $(z_v)_v \longmapsto (b'(z_v))_v$ from $X_A \times X^*_A$ to $\mathfrak{k}_A$ [note that for almost all v we have $b'(z_v) \in o_v$ for $z_v \in X_v^\circ \times (X^*)_v^\circ$].

The adelic topology of $Sp(X)_A$ coincides with the automorphism topology of $Sp(X)_A$ which acts on $X_A \times X_A^*$ by components (this holds, of course, also for char $\mathfrak{k} = 2$): a corresponding result applies in the local case [§ 7.3], and a little reflection will show it for the adelic case once we note that it is enough to consider compact sets in $X_A \times X_A^*$ of the form

$$K = \prod_{v \in P} K_v \times \prod_{v \notin P} (X_v^\circ \times (X^*)_v^\circ) \ , \text{ with } K_v \subset X_v \times X_v^* \text{ compact,}$$

and neighbourhoods U of 0 in $X_A \times X_A^*$ of the form

$$U = \prod_{v \in P} U_v \times \prod_{v \notin P} (X_v^\circ \times (X^*)_v^\circ) \ , \ U_v \text{ a nd. of } 0 \text{ in } X_v \times X_v^* \ ,$$

with the same $P \supset P_\infty$.

We then obtain, using a character χ of $\mathfrak{k}_A$ as in (2-1): the mapping of $Sp(X)_A$ into $Ch_2(X_A \times X_A^*)$ given by $(\sigma_v)_v \longmapsto \chi((b'_{\sigma_v})_v)$ is continuous (analogue of the continuity of $\sigma \longmapsto \psi_\sigma$ in the local case). Further, the mapping $(\sigma_v)_v \longmapsto (b'_{\sigma_v})_v$ of $Sp(X)_A$ into $Q(X)_A$ is continuous (analogue of the continuity of $\sigma \longmapsto f_\sigma$ in the local case).

The proof of (A) can now be given in full analogy to the local case [§ 8.8]. Likewise, the proof of part (B) is analogous to that in § 8.9 (with a few technical details proper to adelization).

The above result shows the significance of the groups $\mathbb{B}(X)_A$, $B_0(X)_A$ when char $\mathfrak{k} \neq 2$.

It is not clear what happens if char $\mathfrak{k} = 2$.

§10. *Weil's theorem 6*

If we combine the representation theorems in §§ 8.6, 9.15 with the general restriction inequality for $\mathfrak{S}^1(G)$ [§ 2.5 (i)], we can obtain Weil's theorem 6 in [25] within the context of Segal algebras.

1. Local case. Let Γ be a closed subgroup of the finite-dimensional vector space X over the local field $\mathbb{K}$. For any $\Phi \in \mathfrak{S}^1(X)$ the integral

$$\underline{S} \longmapsto \int_\Gamma \underline{S}\Phi(\xi)d\xi$$

is a continuous function on Mp(X).

2. Adelic case. Let Γ be a closed subgroup of X_A, the adelization of a finite-dimensional vector space over an 'arithmetic' field [§ 9.1]. For any Φ in the inductive Segal algebra $\mathfrak{S}_I(X_A)$ the integral above is a continuous function on $Mp(X)_A$.

REMARK. These integrals are invariant under left translations by elements of a certain subgroup of Mp(X), respectively $Mp(X)_A$. This follows from Weil's theorem 4 [§ 4.9 above]. Important examples are given by Weil [25, n°s 40, 41]. In Igusa's paper [9, p.221] and in his book [10, Ch. I, § 10] the classical case of $\mathbb{R}^n$ is treated in detail.

3. If the characteristic of the field is not 2, then we may replace in the local case Mp(X) by $\mathbb{B}(X/\mathbb{K})$, and in the adelic case $Mp(X)_A$ by $\mathbb{B}(X)_A$.

The proof is a simple consequence of what we have shown already. The general Segal restriction inequality [§ 2.5 (i)] implies in the local case that

$$\Phi \longmapsto \int_\Gamma \Phi(\xi)d\xi$$

is a continuous linear functional on $\mathfrak{S}^1(X)$. By the local representation theorem [§ 8.6] $\underline{S}\Phi \longrightarrow \underline{S}_0\Phi$ in $\mathfrak{S}^1(X)$ when $\underline{S} \longrightarrow \underline{S}_0$ in Mp(X), whence the result in the local case.

In the adelic case the integral above is, for the same reason, continuous on $\mathfrak{S}^1(X_A)$ and hence on $\mathfrak{S}_I(X_A)$. Using the notation in the adelic representation theorem [§ 9.15], we also have: given $\underline{S}_0 \in Mp(X)_A$ and $\Phi \in \mathfrak{S}_I(X_A)$, there is a P so large that $\underline{S}_0$ lies in the open subgroup $Mp(X)_P^\circ$ and Φ lies in the subspace $\mathfrak{S}_P(X_A)$. We may then take $\underline{S} \in Mp(X)_P^\circ$.

By the representation theorem, $\underline{S}\Phi \longrightarrow \underline{S}_o\Phi$ in $\mathfrak{S}_P(X_A)$, hence also in $\mathfrak{S}_I(X_A)$, when $\underline{S} \longrightarrow \underline{S}_o$ in $Mp(X)^{\bullet}_P$. The result in the adelic case follows.

The assertion for characteristic $\neq 2$ follows in the same way if we apply §§ 8.9, 9.16 (B), respectively.

Thus Weil's main result in [25] has been established in the context of Segal algebras. For this task, many tools of harmonic analysis were used, many difficulties had to be overcome. But it is hoped that a new insight into Weil's result has been gained. The future development of harmonic analysis will, no doubt, see further contributions to this end.

Appendix. Weil's theorem 1 and certain Segal algebras

We prove here the main existence theorem (Weil [25, Th. 1]) for a general l.c.a. group G: given any element $(\sigma,\psi) \in B_0(G)$, there is an operator $S \in \mathcal{B}(G)$ such that $S^{-1}US = U(z\sigma)\psi(z)$, i.e. $\pi_0(S) = (\sigma,\psi)$ (cf. § 4.1 for notation). We shall follow, with some modifications, Weil's proof (which is, in turn, a modification of one given by Segal [22, Th. 2]), and shall emphasize the role of certain Segal algebras in this context.

Let us resume the developments and the notation of § 4.5. We put for $\Phi \in \mathcal{S}^1(G\times G)$

$$(K_\Phi f)(y) := \int_G f(x)\Phi(x,y)dx, \quad f \in \mathcal{K}(G), \; y \in G. \tag{A1}$$

Defining U^v_φ as in § 4 (5-2), (5-3), we now write the relation § 4 (5-4) in the form

$$K_\Phi f = U^v_\varphi f, \text{ with } \Phi(x,y) = \int_{G^*} \varphi(x-y,t^*)\overline{\langle y,t^*\rangle}dt^*, \quad \varphi \in \mathcal{S}^1(G\times G^*), \tag{A2}$$

i.e. $\Phi = \hat{T}^{-1}\varphi$ [cf. § 4 (5-5), (5-8)].

First we show that the operators K_Φ, $\Phi \in \mathcal{S}^1(G\times G)$, form an algebra. Indeed, we have for Φ_1, Φ_2 in $\mathcal{S}^1(G\times G)$

$$K_{\Phi_2}K_{\Phi_1} = K_\Phi, \text{ with } \Phi(x,y) := \int_G \Phi_1(x,u)\Phi_2(u,y)du, \quad (x,y) \in G\times G, \tag{A3}$$

i.e.

$$\Phi := \Phi_1 \times \Phi_2, \tag{A4}$$

the familiar Hilbert-Schmidt product. Note the order on the left-hand side of (A3); this is natural here in view of the definition (A1) which differs slightly from that used by Weil [25, p. 154 (11)]. We show : $\mathcal{S}^1(G\times G)$ is an algebra under the Hilbert-Schmidt product (A4). For this, we write [cf. § 2.6 (ii)]

$$\Phi_1(x,u)\Phi_2(u,y) = \sum_{m\ge1} f_m(x)g_m(u)\cdot \sum_{n\ge1} h_n(u)k_n(y),$$

with f_m, g_m, h_n, k_n in $\mathcal{S}^1(G)$ and

$$\sum_{m\ge1} \|f_m\|_{\mathcal{S}}\cdot\|g_m\|_{\mathcal{S}} < \infty, \quad \sum_{n\ge1} \|h_n\|_{\mathcal{S}}\cdot\|k_n\|_{\mathcal{S}} < \infty.$$

For fixed $m,n \ge 1$ we have

$$|f_m(x)|\cdot\int_G |g_m(u)h_n(u)|du\cdot|k_n(y)| \le |f_m(x)|\cdot\|g_m\|_2\cdot\|h_n\|_2\cdot|k_n(y)| \le$$

$$\le C_1\|f_m\|_6\cdot C_2\|g_m\|_6\cdot C_2\|h_n\|_6\cdot C_1\|k_n\|_6 \quad \text{[cf. § 2 (3-5), (3-6)].}$$

The convergence of $\sum_{m,n\ge 1}\|f_m\|_6\cdot\|g_m\|_6\cdot\|h_n\|_6\cdot\|k_n\|_6$ then shows [cf. (A4)] that

$$\Phi(x,y) = \sum_{m,n\ge 1} c_{mn}f_m(x)k_n(y), \text{ with } c_{mn} = \int_G g_m(u)h_n(u)du,$$

and that $\Phi\in 6^1(G\times G)$. Thus the assertion is proved.

We now consider the operator $U^{\vee}_{\varphi}$ in (A2). By the definition of $U^{\vee}_{\varphi}$ it is readily seen that

$$U^{\vee}_{\varphi_2}U^{\vee}_{\varphi_1} = U^{\vee}_{\varphi} \tag{A5a}$$

with

$$\varphi(u,u^*) := \iint_{GG^*}\varphi_1(u_1,u_1^*)\varphi_2(u-u_1,u^*-u_1^*)\overline{\langle u-u_1,u_1^*\rangle}du_1^*du_1. \tag{A5b}$$

Let us put

$$z_1 := (u_1,u_1^*),\ z := (u,u^*),\ F'(z_1,z) := F(z,z_1)^{-1} \text{ [cf. § 1 (8-2)].}$$

Here F' is again a bicharacter on $(G\times G^*)\times(G\times G^*)$; hence it is a multiplier ('cocycle') for $G\times G^*$. The expression for φ in (A5) is simply the 'twisted convolution' relative to F' (cf. e.g. [18, § 2], where also some references are given):

$$\varphi(z) := \int_{G\times G^*}\varphi_1(z_1)\varphi_2(z-z_1)F'(z_1,z-z_1)dz_1,\quad z\in G\times G^*, \tag{A6a}$$

or, in the usual notation,

$$\varphi = \varphi_1 *_{F'} \varphi_2,\quad \varphi_1,\ \varphi_2 \text{ in } 6^1(G\times G^*). \tag{A6b}$$

Thus the function φ in (A5) may be written in the form (A6). The order on the left-hand side of (A5a) corresponds to that in (A3).

We may, of course, define the twisted convolution (A6) also for φ_1, φ_2 in $L^1(G\times G^*)$ which is then again a Banach algebra, $L^1(G\times G^*,F')$ [cf. loc. cit.]. If we use here F^{-1} instead, we obtain an anti-isomorphic algebra.

We show next: let φ_1, φ_2 be in $6^1(G\times G^*)$; then so is $\varphi_1 *_{F'}\varphi_2$. Thus, with an obvious notation: $6^1(G\times G^*,F')$ <u>is a subalgebra of the twisted convolution algebra</u> $L^1(G\times G^*,F')$. Moreover, <u>the mapping</u> $\hat{T}^{-1}$ [§ 4 (5-5), (5-8)] <u>yields an algebraic isomorphism of</u> $6^1(G\times G^*)$, <u>provided with the twisted convolution</u> $*_{F'}$, <u>onto</u> $6^1(G\times G)$, <u>provided with the Hilbert-Schmidt product</u>:

$$\hat{T}^{-1}(\varphi_1 *_{F'}\varphi_2) = (\hat{T}^{-1}\varphi_1)\times(\hat{T}^{-1}\varphi_2). \tag{A7}$$

To prove this, we evaluate the left-hand term in (A7) directly. Of course, a comparison of (A2)-(A4) with (A5), (A6) furnishes some heuristic argument.

(i) First we note that φ, defined by (A6), is certainly a <u>continuous</u> function in $L^1(G\times G^*)$. This can readily be seen from the fact that the functions φ_j in $\mathcal{S}^1(G\times G^*)$, $j = 1,2$, are, in particular, bounded continuous functions in $L^1(G\times G^*)$.

(ii) Next we show that <u>the function</u> $u^* \longmapsto \varphi(u,u^*)$, $u^*\in G^*$, <u>is integrable over</u> G^* <u>for each</u> $u\in G$, <u>and hence so is</u>

$$u^* \longmapsto \varphi(x-y,u^*)\overline{\langle y,u^*\rangle},\ u^*\in G^*,\ \text{for each } (x,y)\in G\times G.$$

Indeed, the first of these functions is continuous, by (i), and from (A5b) we have, by a permissible interchange of integrals, that

$$\int_{G^*}|\varphi(u,u^*)|du^* \le \iint_{GG^*}|\varphi_1(u_1,u_1^*)|du_1^*du_1\cdot\sup_{v\in G}\int_{G^*}|\varphi_2(v,u^*)|du^*,\ u\in G.$$

The supremum is finite by the restriction inequality § 2 (5-1), applied to $\mathcal{S}^1(G\times G^*)$, the subgroup $\{0\}\times G^*$, and the translated functions $(u,u^*) \longmapsto \varphi_2(u+v,u^*)$.

(iii) For φ as in (A5b) it is thus permissible to evaluate

$$\int_{G^*}\varphi(x-y,u^*)\overline{\langle y,u^*\rangle}du^*$$

by the same interchange of integrals as in (ii). We then obtain a triple integral which reduces, after some simplifications, to $\int_G\Phi_1(x,t)\Phi_2(t,y)dt$, with $\Phi_j = \hat{\tau}^{-1}\varphi_j$, $j = 1, 2$. Thus (A7) <u>is proved</u>.

(iv) It remains to show that $\varphi_1 *_{F'} \varphi_2$ lies in $\mathcal{S}^1(G\times G^*)$. We know that $\Phi_j = \hat{\tau}^{-1}\varphi_j$ is in $\mathcal{S}^1(G\times G)$, hence so is $\Phi_1\times\Phi_2$, as we also know. Since $\hat{\tau}^{-1}$ is a bijection of $\mathcal{S}^1(G\times G^*)$ onto $\mathcal{S}^1(G\times G)$ [§ 4.5], there is a φ' in $\mathcal{S}^1(G\times G^*)$ such that

$$\hat{\tau}^{-1}\varphi' = (\hat{\tau}^{-1}\varphi_1)\times(\hat{\tau}^{-1}\varphi_2).$$

This means by (iii) that [cf. § 4 (5-5)] $(\mathcal{F}_2\varphi')\circ\tau = (\mathcal{F}_2\varphi)\circ\tau$, or $\mathcal{F}_2\varphi' = \mathcal{F}_2\varphi$. It follows [cf.(i), (ii)] that $\varphi' = \varphi$, which completes the proof.

We now extend relation (A7) to functions φ_1, φ_2 in $L^2(G\times G^*)$ as follows. First, (A7) gives, by a familiar property of the Hilbert-Schmidt product and the L^2-isometry of $\hat{\tau}^{-1}$ [§ 4.5]

$$\|\varphi_1 *_{F'} \varphi_2\|_2 \le \|\varphi_1\|_2\cdot\|\varphi_2\|_2 \text{ for } \varphi_1,\ \varphi_2 \text{ in } \mathcal{S}^1(G\times G^*).$$

$\mathcal{S}^1(G\times G^*)$ is dense in $L^2(G\times G^*)$ [§ 2.3, Remark 2]. Let $(\varphi_{1n})_{n\ge1}$, $(\varphi_{2n})_{n\ge1}$ be sequences in $\mathcal{S}^1(G\times G^*)$ tending to φ_1, φ_2 in $L^2(G\times G^*)$, respectively ($n \longrightarrow \infty$). Then the above inequality shows: $(\varphi_{1n} *_{F'} \varphi_{2n})_{n\geq 1}$ is a Cauchy sequence in $L^2(G\times G^*)$. On the other hand, for general φ', φ'' in $L^2(G\times G^*)$ $\varphi' *_{F'} \varphi''$ can be defined as in (A6) and, as is readily seen, is a continuous function. We also have

$$\|\varphi' *_{F'} \varphi''\|_\infty \leq \|\varphi'\|_2 \cdot \|\varphi''\|_2 \text{ for } \varphi',\ \varphi'' \text{ in } L^2(G\times G^*).$$

This shows that $\varphi_{1n} *_{F'} \varphi_{2n}$ tends pointwise, even uniformly, on $G\times G^*$ to the limit $\varphi_1 *_{F'} \varphi_2$ ($n \longrightarrow \infty$). Thus also $\varphi_{1n} *_{F'} \varphi_{2n} \longrightarrow \varphi_1 *_{F'} \varphi_2$ in $L^2(G\times G^*)$ ($n \longrightarrow \infty$). It follows that (A7) <u>holds for general</u> φ_1, φ_2 <u>in</u> $L^2(G\times G^*)$.

The above argument also shows that $L^2(G\times G^*)$ <u>is itself a twisted convolution algebra</u> $L^2(G\times G^*, F')$ and that

$$\|\varphi' *_{F'} \varphi''\|_2 \leq \|\varphi'\|_2 \cdot \|\varphi''\|_2 \text{ for } \varphi',\ \varphi'' \text{ in } L^2(G\times G^*).$$

We are now ready for the proof of the existence theorem stated at the beginning. In $L^2(G\times G^*)$ there is a natural unitary representation r' of $B_0(G)$, given by [cf. § 4 (1-6)]

$$\varphi \longrightarrow r'\varphi\colon\ [r'(\sigma,\psi)\varphi](z) := \varphi(z\sigma)\psi(z),\ z\in G\times G^*\ (\varphi\in L^2(G\times G^*)). \tag{A8}$$

The representation (A8) has the following remarkable property:

$$r'(\varphi_1 *_{F'} \varphi_2) = (r'\varphi_1) *_{F'} (r'\varphi_2),\ r' = r'(\sigma,\psi),\ \varphi_1,\ \varphi_2 \text{ in } L^2(G\times G^*). \tag{A9}$$

Indeed, if we replace in (A6) the functions $\varphi_j \in L^2(G\times G^*)$ by $r'\varphi_j$, $j = 1, 2$, then (A9) appears as a simple consequence of the addition formula for ψ [§ 4 (1-3)] if we write it in the form

$$\psi(z)F'(z_1\sigma,\ z\sigma - z_1\sigma) = \psi(z_1)\psi(z-z_1)F'(z_1, z-z_1),$$

and of the fact that σ has Haar modulus 1 [§ 4.2, Remark 2].

The relation (A9) also holds, of course, in $L^1(G\times G^*, F')$.

The representation (A8) induces a representation r_2 in $L^2(G*G)$ defined by

$$r_2 := \hat{T}^{-1} r' \hat{T},\ r_2 = r_2(\sigma,\psi). \tag{A10}$$

Then (A9), (A10) and (A7) - in its general form - show that

$$r_2(\Phi_1 \times \Phi_2) = r_2(\Phi_1) \times r_2(\Phi_2),\ \Phi_1,\ \Phi_2 \text{ in } L^2(G\times G). \tag{A11}$$

The further proof depends upon the following result:

LEMMA (Weil [25, p. 155, Lemme 3]). <u>The automorphisms</u> r_2 <u>of</u> $L^2(G\times G)$ <u>which have the property</u> (A11) <u>above are precisely those of the form</u>

$$r_2 = S \otimes \bar{S},\ S \in \mathrm{Aut}(L^2(G)), \tag{A12}$$

<u>where</u> $\bar{S} \in \mathrm{Aut}(L^2(G))$ <u>is defined by</u> $\bar{S}g := (S\bar{g})^-$, $g \in L^2(G)$.

The relation (A12) means that for any finite sum $\sum_{n\geq 1} f_n \otimes g_n$ with f_n, g_n in $L^2(G)$ we have $r_2(\sum_{n\geq 1} f_n \otimes g_n) = \sum_{n\geq 1} Sf_n \otimes \bar{S} g_n$.

It is readily seen that, if r_2 has the form (A12), then (A11) holds. Let us prove the converse. First we note that, if $K \in L^2(G\times G)$ satisfies for every $K' \in L^2(G\times G)$ the relation $K\times K'\times K = c_{K'}\cdot K$, with some constant $c_{K'}$, then

$$K = P\otimes Q, \text{ with } P,\ Q \text{ in } L^2(G), \tag{A13}$$

and conversely. To see this, let $K \neq 0$ and take $K' := \Phi'\otimes\Phi''$ with Φ', Φ'' in $L^2(G)$ such that $\iint_{GG} K(x,y)\Phi'(x)\Phi''(y)dxdy \neq 0$. Then we clearly have $K\times K'\times K = \Phi_1\otimes\Phi_2$ with Φ_1, Φ_2 in $L^2(G)$ and $\neq 0$, so $\Phi_1\otimes\Phi_2 = c\cdot K$ with $c \neq 0$, whence (A13). The converse statement is trivial. From this <u>we obtain for</u> P, Q <u>in</u> $L^2(G)$: $r_2(P\otimes Q) = P'\otimes Q'$, <u>with</u> P', Q' <u>in</u> $L^2(G)$. Now choose $P_0 \in L^2(G)$ with $\|P_0\|_2 = 1$ and consider $P_0\otimes\bar{P}_0$: we can write

$$r_2(P_0\otimes\bar{P}_0) = P'_0\otimes Q'_0, \text{ with } \|P'_0\|_2 = 1,\ \|Q'_0\|_2 = 1,$$

since $\|P'_0\|_2\cdot\|Q'_0\|_2 = 1$. Next consider $P\otimes\bar{P}_0$ for any P in $L^2(G)$: we have $P\otimes\bar{P}_0 = (P\otimes\bar{P}_0)\times(P_0\otimes\bar{P}_0)$, hence if $r_2(P\otimes\bar{P}_0) = P_1\otimes Q_1$, say, then we have $r_2(P\otimes\bar{P}_0) = (P_1\otimes Q_1)\times(P'_0\otimes Q'_0) = (P'_0|\bar{Q}_1)\cdot P_1\otimes Q'_0$, with the standard notation $(|)$ for the scalar product in $L^2(G)$. So we can also write, with Q'_0 as above,

$$r_2(P\otimes\bar{P}_0) = P'\otimes Q'_0,\ P' \in L^2(G). \tag{A14a}$$

Likewise we see that $r_2(P_0\otimes Q)$ can be written, with P'_0 as above,

$$r_2(P_0\otimes Q) = P'_0\otimes Q',\ Q' \in L^2(G). \tag{A14b}$$

Moreover, P' in (A14a), Q' in (A14b) are clearly uniquely determined in $L^2(G)$, i.e. they are given by mappings

$$P' = S_1P,\ Q' = S_2Q,$$

which are clearly linear transformations of $L^2(G)$ into $L^2(G)$; also, they preserve the norm, since r_2 preserves the norm in $L^2(G\times G)$ and P_0, Q'_0, P'_0 have norm 1 in $L^2(G)$. Combining (A14) with the obvious relation $P\otimes Q = (P\otimes\bar{P}_0)\times(P_0\otimes Q)$ for P, Q in $L^2(G)$, we obtain

$$r_2(P\otimes Q) = (S_1P_0|[S_2\bar{P}_0]^-)\cdot S_1P\otimes S_2Q.$$

For $P = P_0$, $Q = \bar{P}_0$ this gives $(S_1P_0|[S_2\bar{P}_0]^-) = 1$, hence

$$r_2(P\otimes Q) = S_1P\otimes S_2Q \text{ for all P, Q in } L^2(G). \tag{A15}$$

The same reasoning applies to r_2^{-1}, so S_1, S_2 are invertible. Thus S_1, S_2 are automorphisms of $L^2(G)$. Finally the relation

$$(P \circledast Q) \times (P \circledast Q) = (P|\bar{Q}) \cdot P \circledast Q \text{ for } P, Q \text{ in } L^2(G)$$

shows, if we replace P by S_1P, Q by S_2Q, that [cf. (A15)]

$$(S_1P|\overline{S_2Q}) = (P|\bar{Q}) \text{ or } (S_2Q|\overline{S_1P}) = (Q|\bar{P}) = (S_2Q|S_2\bar{P}).$$

Thus $\overline{S_1P} = S_2\bar{P}$ or $S_2P = (S_1\bar{P})^-$ for $P \in L^2(G)$. Hence by (A15) we obtain (A12), with $S = S_1$, and the lemma is proved.

To prove the theorem, we now apply the lemma to the representation $r_2 = r_2(\sigma, \psi)$ defined by (A8) and (A10): thus there is an $S \in \mathrm{Aut}(L^2(G))$ such that $r_2(f \circledast g) = Sf \circledast \bar{S}g$ for f, g in $L^2(G)$. Replacing g by $\bar{g}$, we obtain $r_2(f \circledast \bar{g}) = Sf \circledast (Sg)^-$. By the very definition of r_2 this gives, if we go back to r' [cf. (A10)]: $r'[\hat{T}(f \circledast \bar{g}) = \hat{T}(Sf \circledast [Sg]^-)$. Now, in view of § 4 (5-9) and the definition of r' [(A8)], the last relation says that

$$(U(z\sigma)\psi(z)f|g) = (U(z)Sf|Sg), \quad z \in G \times G^*.$$

Let f be fixed. Since $g \in L^2(G)$ was arbitrary, it follows that $U(z\sigma)\psi(z)f = S^{-1}U(z)Sf$ or, since f can be any function in $L^2(G)$,

$$U(z\sigma)\psi(z) = S^{-1}U(z)S, \quad z \in G \times G^*,$$

which proves the theorem.

The above proof is based on a natural representation r' of $B_0(G)$ in $L^2(G \times G^*)$. There is a close connexion between this representation and Weil's method of theta representations [§ 4.7]. Indeed, let us consider the group G×G and the subgroup $\Gamma := \{(t,t) | t \in G\}$. Then the group Q of § 4 (7-3) is nothing but G×G*. It turns out that <u>the corresponding theta representation is equivalent to the above representation</u> r' <u>of</u> $B_0(G)$ <u>in</u> $L^2(G \times G^*)$, <u>and the corresponding theta transform is, essentially, the</u> $\hat{T}$<u>-transform</u> used in the preceding proof. This connexion was shown in [20, § III], where the details of the proof can be found.

The proof given here is essentially Weil's modification of that given by Segal [22, Th. 2] in the case when $x \longmapsto 2x$ is an automorphism of G; it yields some novel features of the Segal algebras $\mathfrak{S}^1(G \times G)$ and $\mathfrak{S}^1(G \times G^*)$. The transformation $\hat{T}$ on which the proof rests, also forms the basis of the automorphism theorem [§ 4.5] and is related to a theta transform, as was mentioned above.

The theorem can also be obtained as a by-product of Mackey's famous generalization of the Stone-von Neumann theorem and his theory of projective representations. In this respect, Mackey's work [12,13,14] and the references given there may be consulted.

Bibliography

1. BOURBAKI, N.: Algèbre Commutative, Chaps. 5-6. Paris, Hermann, 1964.
2. —— Espaces Vectoriels Topologiques, Chaps. 1-2. 2nd edn., Paris, Hermann, 1966.
3. —— Intégration, Chaps. 1-4. 2nd edn., Paris, Hermann, 1965.
4. BRACONNIER, J.: Sur les groupes topologiques localement compacts. J. Math. Pures Appl. (9) 27, 1-85 (1948).
5. BURGER, M.: A propos d'un théorème d'André Weil. Travail de diplôme, Institut de mathématiques, Université de Lausanne, 1983 (unpublished).
6. CASSELS, J.W.S.-FRÖHLICH, A. (Eds.): Algebraic Number Theory. London-New York, Academic Press, 1967.
7. FEICHTINGER, H.G.: On a new Segal algebra. Monatsh. Math. 92, 269-289 (1981).
8. FREITAG, E.: Siegelsche Modulfunktionen. Grundlehren d. math. Wiss. 254. Berlin-Heidelberg-New York, Springer, 1983.
9. IGUSA, J.: Harmonic analysis and theta functions. Acta Math. 120, 187-222 (1968).
10. —— Theta Functions. Grundlehren d. math. Wiss. 194. Berlin-Heidelberg-New York, Springer, 1972.
11. LOSERT, V.: A characterization of the minimal strongly character invariant Segal algebra. Ann. Inst. Fourier Grenoble 30, 129-139 (1980).
12. MACKEY, G.W.: Some remarks on symplectic automorphisms. Bull. Amer. Math. Soc. 16, 393-397 (1965).
13. —— Review of Weil's paper [25]. Review 2324. Math. Rev. 29, 448-450 (1965).
14. —— Induced representations of locally compact groups and applications. Functional Analysis and Related Fields (F.E. Brouwer, Ed.), 132-166. Berlin-Heidelberg-New York, Springer, 1970.
15. POGUNTKE, E.: Gewisse Segalsche Algebren auf lokalkompakten Gruppen. Arch. Math. 33, 454-460 (1979).
16. REITER, H.: Classical Harmonic Analysis and Locally Compact Groups. Oxford University Press, 1968.
17. —— L^1-algebras and Segal Algebras. Lecture Notes in Mathematics 231. Berlin-Heidelberg-New York, Springer, 1971.
18. —— Über den Satz von Wiener und lokalkompakte Gruppen. Comment. Math. Helv. 49, 333-364 (1974).
19. —— Über den Satz von Weil-Cartier. Monatsh. Math. 86, 13-62 (1978).
20. —— Theta functions and symplectic groups. Monatsh. Math. 97, 219-232 (1984).
21. ROBERTSON, A.P. & W.J.: Topological Vector Spaces. Cambridge Tracts in Mathematics. 2nd. edn., Cambridge University Press, 1973.

22. SEGAL, I.E.: Transforms for operators and symplectic automorphisms over locally compact abelian groups. Math. Scand. 13, 31-43 (1963).

23. SERRE, J.-P.: Local Fields. Graduate Texts in Mathematics 67. New York-Heidelberg-Berlin, Springer, 1979.

24. SIEGEL, C.L.: Vorlesungen über ausgewählte Kapitel der Funktionentheorie, Bd. I-III. Göttingen, Math. Inst. d. Univ., 1964-1966.

25. WEIL, A.: Sur certains groupes d'opérateurs unitaires. Acta Math. 111, 143-211 (1964) (= [27], vol. III, 1-69).

26. —— Basic Number Theory. Grundlehren d. math. Wiss. 144. 2nd, edn., New York-Heidelberg-Berlin, Springer, 1973.

27. —— Collected Papers, vol. I-III. Corrected second printing. Berlin-Heidelberg-New York, Springer, 1980.

28. WITT, E.: Eine Identität zwischen Modulfunktionen zweiten Grades. Abh. Math. Sem. Univ. Hamburg 14, 323-337 (1941).

Index

Principal notations

LECTURE NOTES IN MATHEMATICS

Edited by A. Dold and B. Eckmann

Some general remarks on the publication of monographs and seminars

In what follows all references to monographs, are applicable also to multiauthorship volumes such as seminar notes.

§1. Lecture Notes aim to report new developments - quickly, informally, and at a high level. Monograph manuscripts should be reasonably self-contained and rounded off. Thus they may, and often will, present not only results of the author but also related work by other people. Furthermore, the manuscripts should provide sufficient motivation, examples and applications. This clearly distinguishes Lecture Notes manuscripts from journal articles which normally are very concise. Articles intended for a journal but too long to be accepted by most journals, usually do not have this "lecture notes" character. For similar reasons it is unusual for Ph.D. theses to be accepted for the Lecture Notes series.

Experience has shown that English language manuscripts achieve a much wider distribution.

§2. Manuscripts or plans for Lecture Notes volumes should be submitted either to one of the series editors or to Springer-Verlag, Heidelberg. These proposals are then refereed. A final decision concerning publication can only be made on the basis of the complete manuscripts, but a preliminary decision can usually be based on partial information: a fairly detailed outline describing the planned contents of each chapter, and an indication of the estimated length, a bibliography, and one or two sample chapters - or a first draft of the manuscript. The editors will try to make the preliminary decision as definite as they can on the basis of the available information.

§3. Lecture Notes are printed by photo-offset from typed copy delivered in camera-ready form by the authors. Springer-Verlag provides technical instructions for the preparation of manuscripts, and will also, on request, supply special staionery on which the prescribed typing area is outlined. Careful preparation of the manuscripts will help keep production time short and ensure satisfactory appearance of the finished book. Running titles are not required; if however they are considered necessary, they should be uniform in appearance. We generally advise authors not to start having their final manuscripts specially tpyed beforehand. For professionally typed manuscripts, prepared on the special stationery according to our instructions, Springer-Verlag will, if necessary, contribute towards the typing costs at a fixed rate.

The actual production of a Lecture Notes volume takes 6-8 weeks.

.../...

§4. Final manuscripts should contain at least 100 pages of mathematical text and should include
- a table of contents
- an informative introduction, perhaps with some historical remarks. It should be accessible to a reader not particularly familiar with the topic treated.
- a subject index; this is almost always genuinely helpful for the reader.

§5. Authors receive a total of 50 free copies of their volume, but no royalties. They are entitled to purchase further copies of their book for their personal use at a discount of 33.3 %, other Springer mathematics books at a discount of 20 % directly from Springer-Verlag.

Commitment to publish is made by letter of intent rather than by signing a formal contract. Springer-Verlag secures the copyright for each volume.

Vol. 1232: P.C. Schuur, Asymptotic Analysis of Soliton Problems. VIII, 180 pages. 1986.

Vol. 1233: Stability Problems for Stochastic Models. Proceedings, 1985. Edited by V.V. Kalashnikov, B. Penkov and V.M. Zolotarev. VI, 223 pages. 1986.

Vol. 1234: Combinatoire énumérative. Proceedings, 1985. Edité par G. Labelle et P. Leroux. XIV, 387 pages. 1986.

Vol. 1235: Séminaire de Théorie du Potentiel, Paris, No. 8. Directeurs: M. Brelot, G. Choquet et J. Deny. Rédacteurs: F. Hirsch et G. Mokobodzki. III, 209 pages. 1987.

Vol. 1236: Stochastic Partial Differential Equations and Applications. Proceedings, 1985. Edited by G. Da Prato and L. Tubaro. V, 257 pages. 1987.

Vol. 1237: Rational Approximation and its Applications in Mathematics and Physics. Proceedings, 1985. Edited by J. Gilewicz, M. Pindor and W. Siemaszko. XII, 350 pages. 1987.

Vol. 1238: M. Holz, K.-P. Podewski and K. Steffens, Injective Choice Functions. VI, 183 pages. 1987.

Vol. 1239: P. Vojta, Diophantine Approximations and Value Distribution Theory. X, 132 pages. 1987.

Vol. 1240: Number Theory, New York 1984–85. Seminar. Edited by D.V. Chudnovsky, G.V. Chudnovsky, H. Cohn and M.B. Nathanson. V, 324 pages. 1987.

Vol. 1241: L. Gårding, Singularities in Linear Wave Propagation. III, 125 pages. 1987.

Vol. 1242: Functional Analysis II, with Contributions by J. Hoffmann-Jørgensen et al. Edited by S. Kurepa, H. Kraljević and D. Butković. VII, 432 pages. 1987.

Vol. 1243: Non Commutative Harmonic Analysis and Lie Groups. Proceedings, 1985. Edited by J. Carmona, P. Delorme and M. Vergne. V, 309 pages. 1987.

Vol. 1244: W. Müller, Manifolds with Cusps of Rank One. XI, 158 pages. 1987.

Vol. 1245: S. Rallis, L-Functions and the Oscillator Representation. XVI, 239 pages. 1987.

Vol. 1246: Hodge Theory. Proceedings, 1985. Edited by E. Cattani, F. Guillén, A. Kaplan and F. Puerta. VII, 175 pages. 1987.

Vol. 1247: Séminaire de Probabilités XXI. Proceedings. Edité par J. Azéma, P.A. Meyer et M. Yor. IV, 579 pages. 1987.

Vol. 1248: Nonlinear Semigroups, Partial Differential Equations and Attractors. Proceedings, 1985. Edited by T.L. Gill and W.W. Zachary. IX, 185 pages. 1987.

Vol. 1249: I. van den Berg, Nonstandard Asymptotic Analysis. IX, 187 pages. 1987.

Vol. 1250: Stochastic Processes – Mathematics and Physics II. Proceedings 1985. Edited by S. Albeverio, Ph. Blanchard and L. Streit. VI, 359 pages. 1987.

Vol. 1251: Differential Geometric Methods in Mathematical Physics. Proceedings, 1985. Edited by P.L. García and A. Pérez-Rendón. VII, 300 pages. 1987.

Vol. 1252: T. Kaise, Représentations de Weil et GL_2 Algèbres de division et GL_n. VII, 203 pages. 1987.

Vol. 1253: J. Fischer, An Approach to the Selberg Trace Formula via the Selberg Zeta-Function. III, 184 pages. 1987.

Vol. 1254: S. Gelbart, I. Piatetski-Shapiro, S. Rallis. Explicit Constructions of Automorphic L-Functions. VI, 152 pages. 1987.

Vol. 1255: Differential Geometry and Differential Equations. Proceedings, 1985. Edited by C. Gu, M. Berger and R.L. Bryant. XII, 243 pages. 1987.

Vol. 1256: Pseudo-Differential Operators. Proceedings, 1986. Edited by H.O. Cordes, B. Gramsch and H. Widom. X, 479 pages. 1987.

Vol. 1257: X. Wang, On the C*-Algebras of Foliations in the Plane. V, 165 pages. 1987.

Vol. 1258: J. Weidmann, Spectral Theory of Ordinary Differential Operators. VI, 303 pages. 1987.

Vol. 1259: F. Cano Torres, Desingularization Strategies for Three-Dimensional Vector Fields. IX, 189 pages. 1987.

Vol. 1260: N.H. Pavel, Nonlinear Evolution Operators and Semigroups. VI, 285 pages. 1987.

Vol. 1261: H. Abels, Finite Presentability of S-Arithmetic Groups. Compact Presentability of Solvable Groups. VI, 178 pages. 1987.

Vol. 1262: E. Hlawka (Hrsg.), Zahlentheoretische Analysis II. Seminar, 1984–86. V, 158 Seiten. 1987.

Vol. 1263: V.L. Hansen (Ed.), Differential Geometry. Proceedings, 1985. XI, 288 pages. 1987.

Vol. 1264: Wu Wen-tsün, Rational Homotopy Type. VIII, 219 pages. 1987.

Vol. 1265: W. Van Assche, Asymptotics for Orthogonal Polynomials. VI, 201 pages. 1987.

Vol. 1266: F. Ghione, C. Peskine, E. Sernesi (Eds.), Space Curves. Proceedings, 1985. VI, 272 pages. 1987.

Vol. 1267: J. Lindenstrauss, V.D. Milman (Eds.), Geometrical Aspects of Functional Analysis. Seminar. VII, 212 pages. 1987.

Vol. 1268: S.G. Krantz (Ed.), Complex Analysis. Seminar, 1986. VII, 195 pages. 1987.

Vol. 1269: M. Shiota, Nash Manifolds. VI, 223 pages. 1987.

Vol. 1270: C. Carasso, P.-A. Raviart, D. Serre (Eds.), Nonlinear Hyperbolic Problems. Proceedings, 1986. XV, 341 pages. 1987.

Vol. 1271: A.M. Cohen, W.H. Hesselink, W.L.J. van der Kallen, J.R. Strooker (Eds.), Algebraic Groups Utrecht 1986. Proceedings. XII, 284 pages. 1987.

Vol. 1272: M.S. Livšic, L.L. Waksman, Commuting Nonselfadjoint Operators in Hilbert Space. III, 115 pages. 1987.

Vol. 1273: G.-M. Greuel, G. Trautmann (Eds.), Singularities, Representation of Algebras, and Vector Bundles. Proceedings, 1985. XIV, 383 pages. 1987.

Vol. 1274: N. C. Phillips, Equivariant K-Theory and Freeness of Group Actions on C*-Algebras. VIII, 371 pages. 1987.

Vol. 1275: C.A. Berenstein (Ed.), Complex Analysis I. Proceedings, 1985–86. XV, 331 pages. 1987.

Vol. 1276: C.A. Berenstein (Ed.), Complex Analysis II. Proceedings, 1985–86. IX, 320 pages. 1987.

Vol. 1277: C.A. Berenstein (Ed.), Complex Analysis III. Proceedings, 1985–86. X, 350 pages. 1987.

Vol. 1278: S.S. Koh (Ed.), Invariant Theory. Proceedings, 1985. V, 102 pages. 1987.

Vol. 1279: D. Ieşan, Saint-Venant's Problem. VIII, 162 Seiten. 1987.

Vol. 1280: E. Neher, Jordan Triple Systems by the Grid Approach. XII, 193 pages. 1987.

Vol. 1281: O.H. Kegel, F. Menegazzo, G. Zacher (Eds.), Group Theory. Proceedings, 1986. VII, 179 pages. 1987.

Vol. 1282: D.E. Handelman, Positive Polynomials, Convex Integral Polytopes, and a Random Walk Problem. XI, 136 pages. 1987.

Vol. 1283: S. Mardešić, J. Segal (Eds.), Geometric Topology and Shape Theory. Proceedings, 1986. V, 261 pages. 1987.

Vol. 1284: B.H. Matzat, Konstruktive Galoistheorie. X, 286 pages. 1987.

Vol. 1285: I.W. Knowles, Y. Saitō (Eds.), Differential Equations and Mathematical Physics. Proceedings, 1986. XVI, 499 pages. 1987.

Vol. 1286: H.R. Miller, D.C. Ravenel (Eds.), Algebraic Topology. Proceedings, 1986. VII, 341 pages. 1987.

Vol. 1287: E.B. Saff (Ed.), Approximation Theory, Tampa. Proceedings, 1985–1986. V, 228 pages. 1987.

Vol. 1288: Yu. L. Rodin, Generalized Analytic Functions on Riemann Surfaces. V, 128 pages, 1987.

Vol. 1289: Yu. I. Manin (Ed.), K-Theory, Arithmetic and Geometry. Seminar, 1984–1986. V, 399 pages. 1987.

Vol. 1290: G. Wüstholz (Ed.), Diophantine Approximation and Transcendence Theory. Seminar, 1985. V, 243 pages. 1987.

Vol. 1291: C. Mœglin, M.-F. Vignéras, J.-L. Waldspurger, Correspondances de Howe sur un Corps p-adique. VII, 163 pages. 1987

Vol. 1292: J.T. Baldwin (Ed.), Classification Theory. Proceedings, 1985. VI, 500 pages. 1987.

Vol. 1293: W. Ebeling, The Monodromy Groups of Isolated Singularities of Complete Intersections. XIV, 153 pages. 1987.

Vol. 1294: M. Queffélec, Substitution Dynamical Systems – Spectral Analysis. XIII, 240 pages. 1987.

Vol. 1295: P. Lelong, P. Dolbeault, H. Skoda (Réd.), Séminaire d'Analyse P. Lelong – P. Dolbeault – H. Skoda. Seminar, 1985/1986. VII, 283 pages. 1987.

Vol. 1296: M.-P. Malliavin (Ed.), Séminaire d'Algèbre Paul Dubreil et Marie-Paule Malliavin. Proceedings, 1986. IV, 324 pages. 1987.

Vol. 1297: Zhu Y.-l., Guo B.-y. (Eds.), Numerical Methods for Partial Differential Equations. Proceedings. XI, 244 pages. 1987.

Vol. 1298: J. Aguadé, R. Kane (Eds.), Algebraic Topology, Barcelona 1986. Proceedings. X, 255 pages. 1987.

Vol. 1299: S. Watanabe, Yu.V. Prokhorov (Eds.), Probability Theory and Mathematical Statistics. Proceedings, 1986. VIII, 589 pages. 1988.

Vol. 1300: G.B. Seligman, Constructions of Lie Algebras and their Modules. VI, 190 pages. 1988.

Vol. 1301: N. Schappacher, Periods of Hecke Characters. XV, 160 pages. 1988.

Vol. 1302: M. Cwikel, J. Peetre, Y. Sagher, H. Wallin (Eds.), Function Spaces and Applications. Proceedings, 1986. VI, 445 pages. 1988.

Vol. 1303: L. Accardi, W. von Waldenfels (Eds.), Quantum Probability and Applications III. Proceedings, 1987. VI, 373 pages. 1988.

Vol. 1304: F.Q. Gouvêa, Arithmetic of p-adic Modular Forms. VIII, 121 pages. 1988.

Vol. 1305: D.S. Lubinsky, E.B. Saff, Strong Asymptotics for Extremal Polynomials Associated with Weights on $\mathbb{R}$. VII, 153 pages. 1988.

Vol. 1306: S.S. Chern (Ed.), Partial Differential Equations. Proceedings, 1986. VI, 294 pages. 1988.

Vol. 1307: T. Murai, A Real Variable Method for the Cauchy Transform, and Analytic Capacity. VIII, 133 pages. 1988.

Vol. 1308: P. Imkeller, Two-Parameter Martingales and Their Quadratic Variation. IV, 177 pages. 1988.

Vol. 1309: B. Fiedler, Global Bifurcation of Periodic Solutions with Symmetry. VIII, 144 pages. 1988.

Vol. 1310: O.A. Laudal, G. Pfister, Local Moduli and Singularities. V, 117 pages. 1988.

Vol. 1311: A. Holme, R. Speiser (Eds.), Algebraic Geometry, Sundance 1986. Proceedings. VI, 320 pages. 1988.

Vol. 1312: N.A. Shirokov, Analytic Functions Smooth up to the Boundary. III, 213 pages. 1988.

Vol. 1313: F. Colonius, Optimal Periodic Control. VI, 177 pages. 1988.

Vol. 1314: A. Futaki, Kähler-Einstein Metrics and Integral Invariants. IV, 140 pages. 1988.

Vol. 1315: R.A. McCoy, I. Ntantu, Topological Properties of Spaces of Continuous Functions. IV, 124 pages. 1988.

Vol. 1316: H. Korezlioglu, A.S. Ustunel (Eds.), Stochastic Analysis and Related Topics. Proceedings, 1986. V, 371 pages. 1988.

Vol. 1317: J. Lindenstrauss, V.D. Milman (Eds.), Geometric Aspects of Functional Analysis. Seminar, 1986–87. VII, 289 pages. 1988.

Vol. 1318: Y. Felix (Ed.), Algebraic Topology – Rational Homotopy. Proceedings, 1986. VIII, 245 pages. 1988

Vol. 1319: M. Vuorinen, Conformal Geometry and Quasiregular Mappings. XIX, 209 pages. 1988.

Vol. 1320: H. Jürgensen, G. Lallement, H.J. Weinert (Eds.), Semigroups, Theory and Applications. Proceedings, 1986. X, 416 pages. 1988.

Vol. 1321: J. Azéma, P.A. Meyer, M. Yor (Eds.), Séminaire de Probabilités XXII. Proceedings. IV, 600 pages. 1988.

Vol. 1322: M. Métivier, S. Watanabe (Eds.), Stochastic Analysis. Proceedings, 1987. VII, 197 pages. 1988.

Vol. 1323: D.R. Anderson, H.J. Munkholm, Boundedly Controlled Topology. XII, 309 pages. 1988.

Vol. 1324: F. Cardoso, D.G. de Figueiredo, R. Iório, O. Lopes (Eds.), Partial Differential Equations. Proceedings, 1986. VIII, 433 pages. 1988.

Vol. 1325: A. Truman, I.M. Davies (Eds.), Stochastic Mechanics and Stochastic Processes. Proceedings, 1986. V, 220 pages. 1988.

Vol. 1326: P.S. Landweber (Ed.), Elliptic Curves and Modular Forms in Algebraic Topology. Proceedings, 1986. V, 224 pages. 1988.

Vol. 1327: W. Bruns, U. Vetter, Determinantal Rings. VII,236 pages. 1988.

Vol. 1328: J.L. Bueso, P. Jara, B. Torrecillas (Eds.), Ring Theory. Proceedings, 1986. IX, 331 pages. 1988.

Vol. 1329: M. Alfaro, J.S. Dehesa, F.J. Marcellan, J.L. Rubio de Francia, J. Vinuesa (Eds.): Orthogonal Polynomials and their Applications. Proceedings, 1986. XV, 334 pages. 1988.

Vol. 1330: A. Ambrosetti, F. Gori, R. Lucchetti (Eds.), Mathematical Economics. Montecatini Terme 1986. Seminar. VII, 137 pages. 1988.

Vol. 1331: R. Bamón, R. Labarca, J. Palis Jr. (Eds.), Dynamical Systems, Valparaiso 1986. Proceedings. VI, 250 pages. 1988.

Vol. 1332: E. Odell, H. Rosenthal (Eds.), Functional Analysis. Proceedings, 1986–87. V, 202 pages. 1988.

Vol. 1333: A.S. Kechris, D.A. Martin, J.R. Steel (Eds.), Cabal Seminar 81–85. Proceedings, 1981–85. V, 224 pages. 1988.

Vol. 1334: Yu.G. Borisovich, Yu. E. Gliklikh (Eds.), Global Analysis – Studies and Applications III. V, 331 pages. 1988.

Vol. 1335: F. Guillén, V. Navarro Aznar, P. Pascual-Gainza, F. Puerta, Hyperrésolutions cubiques et descente cohomologique. XII, 192 pages. 1988.

Vol. 1336: B. Helffer, Semi-Classical Analysis for the Schrödinger Operator and Applications. V, 107 pages. 1988.

Vol. 1337: E. Sernesi (Ed.), Theory of Moduli. Seminar, 1985. VIII, 232 pages. 1988.

Vol. 1338: A.B. Mingarelli, S.G. Halvorsen, Non-Oscillation Domains of Differential Equations with Two Parameters. XI, 109 pages. 1988.

Vol. 1339: T. Sunada (Ed.), Geometry and Analysis of Manifolds. Procedings, 1987. IX, 277 pages. 1988.

Vol. 1340: S. Hildebrandt, D.S. Kinderlehrer, M. Miranda (Eds.), Calculus of Variations and Partial Differential Equations. Proceedings, 1986. IX, 301 pages. 1988.

Vol. 1341: M. Dauge, Elliptic Boundary Value Problems on Corner Domains. VIII, 259 pages. 1988.

Vol. 1342: J.C. Alexander (Ed.), Dynamical Systems. Proceedings, 1986–87. VIII, 726 pages. 1988.

Vol. 1343: H. Ulrich, Fixed Point Theory of Parametrized Equivariant Maps. VII, 147 pages. 1988.

Vol. 1344: J. Král, J. Lukeš, J. Netuka, J. Veselý (Eds.), Potential Theory – Surveys and Problems. Proceedings, 1987. VIII, 271 pages. 1988.

Vol. 1345: X. Gomez-Mont, J. Seade, A. Verjovski (Eds.), Holomorphic Dynamics. Proceedings, 1986. VII, 321 pages. 1988.

Vol. 1346: O. Ya. Viro (Ed.), Topology and Geometry – Rohlin Seminar. XI, 581 pages. 1988.

Vol. 1347: C. Preston, Iterates of Piecewise Monotone Mappings on an Interval. V, 166 pages. 1988.

Vol. 1348: F. Borceux (Ed.), Categorical Algebra and its Applications. Proceedings, 1987. VIII, 375 pages. 1988.

Vol. 1349: E. Novak, Deterministic and Stochastic Error Bounds in Numerical Analysis. V, 113 pages. 1988.